Contents

List of Illustrations ix

List of Maps xii

Acknowledgements xiii

Timeline xv

Family Trees xvii

Introduction 1

Prologue 4

Part I

1 Radicals in Suburbia 9

2 Learning Curves 19

3 Finding Their Own Way 31

4 'Another Word for Suicide' 43

5 'Fellowship is Heaven' 53

Part II

6 Answering the Call 69

7 Not Much 'Home' About It 85

8 Behind Closed Doors 97

9 Girls in Trades 110

10 Medical Men 122

Part III

11 A Place to Begin Again 137

12 'Ignoble Motives' 146

13 The Politics of Knitting 162

14   Landfall   176

15   After Lives   186

Notes   199

Select Bibliography   220

Index   227

List of Illustrations

1.1 Elizabeth Cash with the children in about 1876 (reproduced with the kind permission of James Cash)   10

1.2 Poster advertising the 'fresh air' suburb of Upper Norwood in about 1900 (reproduced with the kind permission of Melvyn Harrison, Chairman, Crystal Palace Foundation)   12

2.1 Harold Oakeshott as a young man (reproduced with the kind permission of Gillian Oxford)   22

2.2 Senior pupils at Croydon High School for Girls c.1888 (reproduced with the kind permission of Croydon High School)   28

2.3 The Upper Sixth Form at Croydon High Schools for Girls in 1891 (reproduced with the kind permission of Croydon High School)   29

3.1 Class of juniors at Croydon High School for Girls (probably in 1891 or 1892) (reproduced with the kind permission of Croydon High School)   37

4.1 Downside Cottage in Fanfare Road (later Downs Road) in Coulsdon, Surrey (reproduced with the kind permission of Gillian Oxford)   48

5.1 Hand traced map (taken from the sailing log for 1899) of the Hampshire, Dorset and Devon coastline (reproduced with the kind permission of James Cash)   62

5.2 Photograph of Walter, Harold and Grace, on the East Coast cruise of August 1900 (reproduced with the kind permission of Cherry Dingemans)   65

6.1 Walter as a small boy (reproduced with the kind permission of Cherry Dingemans)   70

6.2 A view in 1827 of The Church Missionary Society College in Islington, London (© London Metropolitan Archives, City of London)   71

6.3 A photograph, taken in 1901, of an early steamer in the Northwest Territories (NWT Archives, Yellowknife, C. W. Mathers, fonds/N-1979-058:0008)   73

6.4 The pulpit in the modern-day St David's Anglican Church, Fort Simpson (photograph taken by the author in 2011)   75

6.5   William Day Reeve in the 1890s, soon after his consecration
       as the second Bishop of Mackenzie River (reproduced with
       the kind permission of Cherry Dingemans)   84

7.1   A view of the Home in Highbury Grove, Islington, London
       (© London Metropolitan Archives, City of London)   86

7.2   Monkton Combe's First XV in 1892 (reproduced with
       the kind permission of Monkton Combe School)   95

8.1   Working at home: making matchboxes (*c*.1900) (reproduced
       with the kind permission of Bishopsgate Library, Bishopsgate
       Foundation and Institute, London)   103

9.1   A class in waistcoat-making at Borough Polytechnic Institute,
       in the early 1900s (reproduced with the kind permission
       of London South Bank University)   116

9.2   The Stanley Gymnasium at Borough Polytechnic Institute
       in the early 1900s (reproduced with the kind permission
       of London South Bank University)   118

10.1   A view of Guy's Memorial Park, taken from the colonnade
       in 1925 (reproduced with the kind permission of the
       Gordon Museum, King's College London)   125

10.2   The operating theatre at Guy's Hospital, 1890 (reproduced
       with the kind permission of the Gordon Museum,
       King's College London)   127

11.1   Overlooking Wellington City, 1905 (reproduced with
       the kind permission of Alexander Turnbull Library, Wellington,
       New Zealand. Watt, T (Miss), fl 1978: Photographs of
       Wellington and Napier scenes. Ref: 1/2-080439-F)   142

11.2   The fire at Parliament Buildings, Wellington, in 1907
       (reproduced with the kind permission of Alexander
       Turnbull Library, Wellington, New Zealand. Zachariah,
       Joseph, 1867–1965. Lawes, Mr, fl 1961: Photographs.
       Ref: 1/2-019528-F)   144

12.1   Redstone's coach travelling north of Gisborne in about 1910,
       possibly on Makorori beach (reproduced with the kind
       permission of Alexander Turnbull Library, Wellington,
       New Zealand. Williams, F. J. (Mrs), fl 1967: Photographs of
       coaches in the East Coast region. Ref: 1/2-029781-F)   147

12.2   A postcard view of Kaiti taken in about 1910 by an unknown
       photographer (reproduced with the kind permission of
       Alexander Turnbull Library, Wellington, New Zealand. Johnston,
       I. F. H. (Mr): Postcards of New Zealand. Ref: PAColl-6001-18)   150

12.3 A kindergarten class in Gisborne in about 1912 (reproduced with the kind permission of Cherry Dingemans) 152

13.1 Pupils of Cook County College for Girls in front of their school in about 1916 (reproduced by kind permission of the Tairawhiti Museum, Gisborne, New Zealand, Ref. 251–4) 167

13.2 Boys and girls sewing and knitting in the city schools of Auckland (reproduced with the kind permission of Sir George Grey Special Collections, Auckland Libraries, AWNS-19150715-37-1) 169

13.3 The Reeve family with friends in 1918, at Waihuka, near Gisborne (reproduced with the kind permission of Mike Stockwell) 173

14.1 'Ranmore', in Duart Road, Havelock North (photograph taken by the author in 2008) 178

14.2 A street in Havelock North, Hawke's Bay, in about 1925 (reproduced with the kind permission of Alexander Turnbull Library, Wellington, New Zealand. Smith, Sydney Charles, 1888–1972: Photographs of New Zealand. Ref: 1/2-046113-G) 181

14.3 The headstone on Joan's grave in the Havelock North cemetery (photograph taken by the author in 2008) 185

15.1 Harold with Dorothy, at the time of their engagement (reproduced with the kind permission of Gillian Oxford) 188

15.2 Henry Cash on 'Peggy', in about 1948 (reproduced with the kind permission of James Cash) 194

15.3 Harold with his youngest child, David (reproduced with the kind permission of Gillian Oxford) 195

List of Maps

Map of Surrey, 1885 (© Sophie Devine)   8

Map of the Northwest Territories, 1870 (© Sophie Devine)   68

Map of North Island, New Zealand (© Sophie Devine)   136

Acknowledgements

There can be few things more disconcerting than a stranger fetching up on your doorstep knowing more about your family than you do. The families in this story have always responded with grace and generosity to my inquiries and the book could not have been written without them. Sophie Dingemans, Grace's great-granddaughter, was first to let the cat out of the bag (with her fictional play *Grace* performed in New Zealand in 2008) and special thanks are due to her. Her mother, Cherry, has patiently replied to my many questions over several years and warmly welcomed me to stay whenever I was in New Zealand. James and Susan Cash, Gillian Oxford and Marion Shea have also trusted me with their stories, memories, photographs and family papers. Thanks also to Anna Hannam (for leading me to her parents), and to Joanne Jones, Mary Miles, Martin Reeve and Mike Stockwell for their recollections of family life. Robin Dingemans (and other Reeve family descendants), Susan Cash, Priscilla Oakeshott and Claire Hill have kindly contributed findings from their genealogical research.

Thank you also to those who have shared their local knowledge and expertise, including Jim Farrell (for showing me Limpsfield and its environs), Malcolm Gifford (for a trip to see the East Coast rivers) and Margaret Peterson (for taking me to Rae in the Northwest Territories). Ann Robson suggested fruitful avenues of enquiry into the lives and activities of missionaries and indigenous peoples in Canada in the late nineteenth century. Guy Lanoe (Society for the Protection and Development of the Heritage and History of Arzon) supplied background about Arzon, and Urszula Watrobska kept me company in Brittany where we explored the archives and beaches together. The late Margaret Robb kindly translated some historical material and Rachel Singh searched past copies of French newspapers.

In New Zealand, Margaret Shanks and Marie Burgess provided valuable assistance with research in Gisborne, and John Cochran, Elizabeth Cox and Meghan Hughes (all of Wellington) delved into historical records on my behalf and uncovered many significant details. I am also grateful to my good friend, Jan Hope, who drove me to Paeroa. Here, bent double, we scoured a rain-lashed hillside together, looking for Walter's tombstone. Alan and Diana McDonald and Paul Von Dadelszen of Havelock North provided valuable background about the activities of 'Havelock Work' in the early twentieth century.

Margaret Walker accompanied me down to a cold basement in Croydon High School where I found my first photograph of Grace as a young girl. Bill Edwards lent his expertise to the interpretation of some early photographs taken at Guy's Medical School and the late Dr Houston contributed useful observations about the institution's history. John Owen Smith introduced me to Headley Down in Hampshire where Kate Cash spent her final years and Daintry Midgley and Angela Gillon shared their recollections of Renée from the 1950s and early 1960s. Sue Saxby-Smith came with me to Ranmore Common in Surrey where we walked in Grace's footsteps.

Professor Jonathan Reinarz and Professor Margaret Tennant kindly gave expert advice on draft chapters. Thank you also to Kate Chisholm, Merryn Hutchings, Paula Lanning, and especially Bill Bailey, for reading draft chapters and offering many insights and helpful suggestions. I am also indebted to Julie Wheelwright, Sarah Bakewell, Julian Putkowski and Carole Seymour-Jones, as well as fellow students on the narrative non-fiction writing course at City University in London, for their early encouragement and guidance. The City Lit workshops, run by Christina Dunhill, gave me further opportunities to share my writing as it progressed and I am grateful for the feedback and suggestions I received there. Thanks to Clare Mence for her faith in the book, to Emily Russell (her successor at Palgrave) for continuing to believe in it and to Angharad Bishop for her editorial assistance. Thank you also to Sophie Devine for the maps and family trees.

My warm thanks to staff at the following institutions: Eltham College, Croydon High School, Monkton Combe School, Wanganui Collegiate School, Royal Grammar School Newcastle and Woodford House, Havelock North. Thank you, too, to archivists and librarians at the Cadbury Research Library (Birmingham University), London Metropolitan Archives, City of Westminster Archives Centre, the UCL Institute of Education (London), King's College London, Brunel University Archives, London School of Economics, the University of Roehampton, Trades Union Congress Library Collections (London Metropolitan University), University of London (Senate House Library), University College London (Records Office), The National Archives, The Gordon Museum of Pathology (King's College London), The Old Operating Theatre Museum (London), the St Bride Foundation Library (London), the Museum of Croydon, the Croydon Natural History & Scientific Society, Torbay Council, the Institution of Engineering and Technology, and the Electrical Contractors' Association.

In New Zealand I received valuable assistance from staff at the Museum of New Zealand Te Papa Tongawera, the Alexander Turnbull Library, the National Library of New Zealand, and Archives New Zealand, as well as the Tairawhiti Museum (Gisborne), HB Williams Memorial Library (Gisborne), Gisborne District Council, the Department of Conservation (Gisborne Office) and Hauraki District Council (Paeroa). The assistance of staff at the Prince of Wales Northern Heritage Centre in Yellowknife, Canada, is also gratefully acknowledged. Thank you to Copyright Licensing and the New Zealand Society of Authors for the CLNZ/NZSA Research Grant in 2009.

Timeline

1670   The Hudson's Bay Company is formed by English royal charter.

1799   The Church Missionary Society is founded in London.

1840   The Treaty of Waitangi is signed by representatives of the British Crown and various Maori chiefs in New Zealand.

1857   In the UK, the first Matrimonial Causes Act reforms the law on divorce, widens its availability and abolishes adultery as a criminal offence.

1867   Four specifically Maori electorates are established by the New Zealand parliament.

1868   The Schools Inquiry Commission publishes the Taunton Report in England.

1870   The Education Act sets up locally elected School Boards in England and Wales to provide elementary school places for all working-class children.

1871   Newnham College is founded in Cambridge, England.

1872   The Girls' Public Day School Company is established in London.

1877   Free, secular and compulsory primary education is introduced in New Zealand.

1883   The Fellowship of the New Life is formed in London.

1884   The first meeting of the Fabian Society takes place in London.

1888   *The Diary of a Nobody* is first published as a serial in *Punch* magazine.

1889   The first issue of *Seed-time* is published. The first London County Council is elected.

1890   The discovery of an effective antitoxin for diphtheria is announced in Germany. *News from Nowhere* by William Morris first appears in print.

1892   In London, the Froebel Educational Institute is constituted.

1893   Women in New Zealand win the right to vote.

1894   The inaugural meeting of the Women's Industrial Council takes place in London.

1895   X-rays are discovered by a German physicist.

1898   Old age pensions are introduced in New Zealand. In England, the Fellowship of the New Life is disbanded.

1904   The first London Trade School for Girls opens at Borough Polytechnic Institute, specialising in waistcoat-making.

1906   *In the Days of the Comet* by H. G. Wells appears in print. *A Doctor's Dilemma* by George Bernard Shaw is first performed.

1907   New Zealand becomes a dominion. Truby King founds the Plunkett
       Society in Dunedin.
1914   The German army invades Belgium and the First World War begins.
1915   New Zealanders take part in the Gallipoli landing in Turkey.
1917   The Church Missionary Society College in London closes.
1920   Jane Mander publishes *The Story of a New Zealand River*.
1924   Ramsay MacDonald is elected first Labour Prime Minister of the
       United Kingdom.
1928   Alexander Fleming discovers penicillin. Women in the United
       Kingdom achieve the same voting rights as men.
1935   Michael Joseph Savage is elected first Labour Prime Minister of
       New Zealand.
1948   Women are admitted to full membership of Cambridge University,
       England.

Family Trees

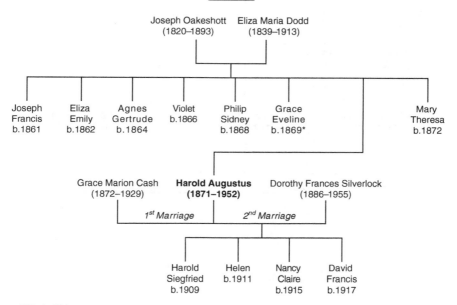

Oakeshott

Joseph Oakeshott   Eliza Maria Dodd
(1820–1893)   (1839–1913)

| Joseph Francis b.1861 | Eliza Emily b.1862 | Agnes Gertrude b.1864 | Violet b.1866 | Philip Sidney b.1868 | Grace Eveline b.1869* | Mary Theresa b.1872 |

Grace Marion Cash   **Harold Augustus**   Dorothy Frances Silverlock
(1872–1929)   **(1871–1952)**   (1886–1955)

1st Marriage   *2nd Marriage*

| Harold Siegfried b.1909 | Helen b.1911 | Nancy Claire b.1915 | David Francis b.1917 |

* Died within a year

xvii

<u>Cash</u>

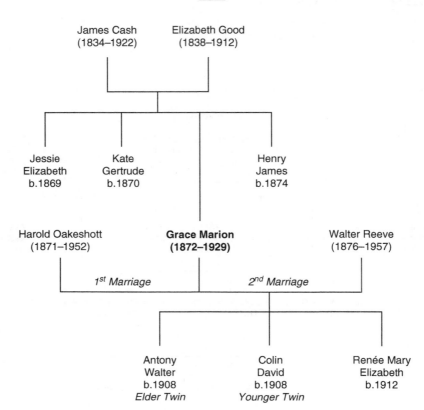

Reeve

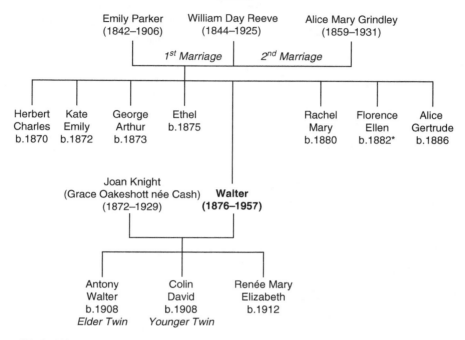

Emily Parker (1842–1906) William Day Reeve (1844–1925) Alice Mary Grindley (1859–1931)

1st Marriage   2nd Marriage

Herbert Charles b.1870 | Kate Emily b.1872 | George Arthur b.1873 | Ethel b.1875 | Rachel Mary b.1880 | Florence Ellen b.1882* | Alice Gertrude b.1886

Joan Knight (Grace Oakeshott née Cash) (1872–1929)   **Walter (1876–1957)**

Antony Walter b.1908 *Elder Twin* | Colin David b.1908 *Younger Twin* | Renée Mary Elizabeth b.1912

*Died within a year

'Would you not like to try *all* sorts of lives – one is so very small.'
Katherine Mansfield in a letter to Sylvia Payne,
dated 24 April 1906,
The Collected Letters of Katherine Mansfield.

Introduction

In the summer of 2008 some old photographs caught my eye. Tied like bundles in their loose dresses, some girls were hanging from thick ropes in a high-ceilinged gymnasium. Others posed awkwardly beside various pieces of apparatus – a rope ladder, a climbing frame, a vaulting horse. In a second photograph, the girls were stooped over needlework tasks, and in another, they were fitting lengths of expensive-looking fabric to dressmaking dummies in a workshop. I was curious.

My background is in vocational education and I was still employed at that time as an academic. I began to investigate. It turned out that the girls in the photographs were all attending the London Trade Schools in the early 1900s, at what was then Borough Polytechnic in south London. I soon discovered that these schools were the inspiration of one Grace Oakeshott and her colleagues at the Women's Industrial Council. So who was she, I wondered? What other contributions had she made to girls' education?

I was disappointed to read that a few years after the Trade Schools began, Grace had drowned in France, at the age of only 35, and for a while my investigations came to a halt.

Back at work, I couldn't quite get Grace out of my mind. Endless googling that summer had yielded nothing new, but now, to my amazement, at the top of the list of retrievals was a review of a play called 'Grace' that had been written by a young woman living in New Zealand. Sophie Dingemans was, she said, the great-granddaughter of a woman called Grace Oakeshott who had staged her own death in 1907 and run away with her lover.

Could this be the same Grace that I was interested in? Certainly her name was spelt the same and the dates fitted. Other details I was less sure of but when repeated searches kept turning up the same review, the same interview with Sophie who was making the same assertions, I resolved to contact her. She put me in touch with her mother, Cherry, and so it began. I never doubted the family's story; they had always known who 'Joan' really was and could

1

document several of their claims. They were keen to find out how much I knew about Grace's life in England. At that stage, I knew very little. The family told me that she had been married to Harold Oakeshott and that he had re-married after her departure. Using what little information I had, I pursued Grace, uncovering her story slowly, building from a detail here or a detail there.

Although I live in London, New Zealand is my home country and coming to know it in a new way has been a joy. As well as several locations there, including Gisborne and Havelock North, the story has taken me to the Northwest Territories in Canada, to Brittany in France, and many places in England – including Thornton Heath, Croydon, Coulsdon, Limpsfield, Ramsgate, Burnham-on-Crouch and Babbacombe Bay on the Devon coast. When I realised I had at last tracked down some descendants of Henry Cash, Grace's brother, and was to meet them the next day in Kent, I was too excited to sleep. And the moment I discovered that one of Harold's granddaughters lived a short distance from me in north London is also one that will stay with me.

During five or six years of research, I rode a rollercoaster. Some of my discoveries confirmed early hunches; others challenged them and threw doubt on previous findings. There were false trails and blind alleys. I spent frustrating hours delving into the lives and activities of people who would turn out to have little or no connection with Grace. Often she was invisible in the places I most expected to find her. Sometimes I focused on academic sources, gathering material about the times she lived in, her professional and political activities, her colleagues and associates; at other times, I pursued leads that the families had given me. Often as I set out, I would imagine discovering hidden stashes of documents. Perhaps I would find a letter written by Grace to her sister, Jessie, composed on a beach in Brittany, or a diary of her marriage to Harold, or her first days in New Zealand. Perhaps I would find a letter written by her lover, Walter, to his mother or his father, the Bishop. But even without such first-hand accounts, it is a story that hooks most people who hear it and early on I resolved not to fictionalise Grace's life or the lives of those around her. It is its truthfulness that fascinates.

This book is written in the style of narrative non-fiction. Dialogue has been constructed from published material in newspapers and journals or family recollections, and details and descriptions of scenes are based on historical material or my personal observations of places that I have visited. The lives of the people in this story have presented themselves with gaps, as most lives do, and not everything is known about Grace, Harold or Walter. They flit through the chapters, sometimes shadowy, sometimes centred and more substantial.

Yet one thing was always clear to me. They were serious people and they believed in using their talents and opportunities to help others. The decades known as *fin de siècle* were exciting and full of uncertainty. It was a time to experiment. Grace's life (like the lives of those around her) has individual interest but it also reflects and offers insights into the preoccupations, thoughts and beliefs of the period. Debates about marriage and the role of women, vocational training for girls, medical advances, colonial imperialism, the alleviation of poverty and the ethics of socialism provided the backdrop to their lives. The reader will travel across continents, from Europe to North America and the Antipodes, tracing the differing passions and activities of the main characters, their families and associates. My intention is to tell the common story through Grace's uncommon life.

About money

Comparing sums of money between the nineteenth century and today can be complicated. Historical wages, for example, only have meaning when the contemporary prices of commodities are known. In the book, some comparisons have been made in relation to wages and salaries, and these are based on calculations by the economists Lawrence Officer and Samuel Williamson on their website www.measuringworth.com. The figures given have been calculated using the website's 'historic standard of living' measure which estimates the purchasing power of an income or wealth for a past year in terms of today's pound sterling.

Prologue

There could be no doubt that the clothes belonged to a woman. The long skirts and petticoats were piled neatly on the deserted beach in Brittany where just days earlier a party of English friends had been seen relaxing together. The clothes were modest and simple, and they suggested a respectable, middle-class lifestyle.

It was Tuesday, 27 August 1907. The skies that day were clear. There was a light northerly wind and the temperature was near normal for the time of year. English visitors often came to Brittany in the holiday season. In the wake of the romantic revival, with its focus on Brittany's Celtic past, they were keen to experience the architecture, painting and language of the region. Most holidaymakers visited the north coast but a few more adventurous ones came south, to small towns and coastal villages, like Arzon. Some visitors wanted to see the archaeological sites where standing stones grew like truffles. In 1853 archaeologists had discovered a prehistoric tomb and a fine collection of grave goods underneath a burial mound known as Caesar's Mound since, according to legend, the Roman general had used it as a vantage point during a naval battle in 56 BC.

Arzon lies towards the tip of the Rhuys Peninsular in the Morbihan Gulf. In the nineteenth century, you could reach it by taking a horse and cart from Vannes, the largest town in the region, but the journey might take you more than four hours. Boatmen provided a ferry service to many destinations within the Gulf, including Arzon, and local inhabitants journeyed from place to place, for a family wedding, a religious ceremony or a market. Tides and winds determined the timetable, and the charges for people (as well as for pigs and cows) were fixed by the municipal councils. By 1900 travel had become easier and a steamer provided a regular service from Vannes to nearby Port Navalo, stopping at all the islands on the way. Arzon was now developing quickly. The little town had recently acquired schools, a council chamber, a fish market, and wharves; a railway would soon be constructed and would provide a reliable link with local settlements. The first hotel, Hôtel de la Plage, was reviving tourism following the failure of a local casino. Swimming was popular with holidaymakers and though their clumsy, woollen 'bathing gowns' would soon be replaced by lighter attire, it was still considered indecent for women to show their arms, legs or necks.

It was not unusual to see small piles of clothes on the beach in Arzon. The town faces the Atlantic and the smaller beaches, tucked between rocky promontories, afforded privacy to bathers. Warm currents lifted them and carried them towards the tiny islands scattered across the bay. But there were cold currents, too, and tides which might catch you unawares. A day later, for example, there were three reports of bodies being recovered from the water,[1] including that of Monsieur Lancelot, whose widow had lost her first husband in the same way. Lancelot's body was recovered a few days after his shoes and some personal effects had been found on the shore.

In England, on 2 September, a death notice appeared in *The Times*. An English woman, Grace Marion Oakeshott (née Cash), the wife of Harold Augustus Oakeshott, had died at Arzon in Brittany, aged only 35. Four days later, an obituary in the same newspaper claimed she had drowned 'whilst swimming'. She was described as 'one of the most capable of the younger workers for the industrial progress of women'.[2] Eight days after that, the Independent Labour Party in Croydon put a notice in the local newspaper:

> Most of our members will have heard by now of the cruel blow sustained by our Comrade H. A. Oakeshott whose wife was drowned while bathing on the Atlantic shore of Brittany. The horror of his loss is not lessened by the fact that there seems small chance of the body being recovered. While the hearts of all his comrades will go out in sympathy to Mr Oakeshott, whose worth and industry in the labour movement we so much appreciate, at the same time we are deeply conscious of the great loss Mrs Oakeshott's untimely end will be to the women of the cause.[3]

In October, *The Englishwoman's Review* published an obituary explaining that Grace Oakeshott had been responsible for the establishment of the first Trade School for Girls in London, for waistcoat-making.[4] And at the end of the year, in a black-rimmed box that appeared in their Thirteenth Annual Report, her colleagues at the Women's Industrial Council placed on record their 'deep sense of the irreparable loss' they had sustained 'in common with all workers amongst industrial women'.[5]

Registry documents in London show that Harold Augustus Oakeshott married again, almost exactly a year after Grace's death, in July 1908. The marriage certificate gave his 'condition' as 'widower'. In May 1909 his new wife, Dorothy, gave birth to a son. In 1911, when they completed the Census form,

Grace's parents, James and Elizabeth Cash, entered the figure '4' in a column headed 'total children born alive', and '3' for 'children still living'. Although a body had not been found, and no death certificate was ever issued, in England at least Grace's death was now a matter of official record.

It is also a matter of record that in October 1907 two people who gave their names as 'Dr and Mrs Reeve' boarded the *Orotava* in Marseilles and travelled to Sydney (via Freemantle) in Australia. They gave their ages as 31 and 34 respectively. A few days later, in Sydney, a 'Mr and Mrs Reeve' boarded the *Manuka* and voyaged across the Tasman Sea to New Zealand. They left the ship in Wellington. On the passenger list this time, the age difference was reversed. The man gave his as 28 and his companion's age was recorded as 25. But they were the same couple that had joined the *Orotava* in Marseilles a few weeks before – 'Mr Reeve' was Walter, the son of an Anglican missionary, and 'Mrs Reeve' was Grace Marion Oakeshott, embarking on a new life in a new country, with her lover.

Part I

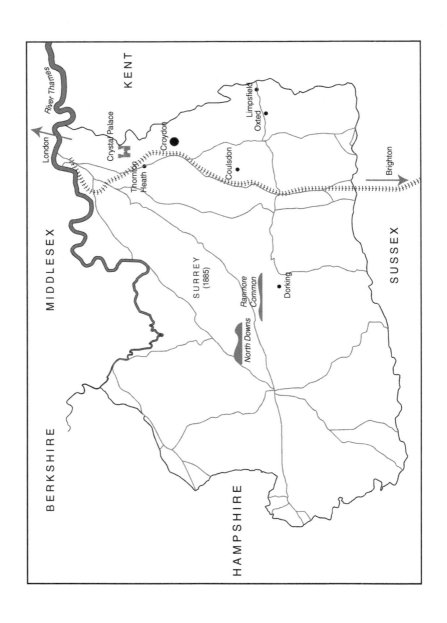

1
Radicals in Suburbia

The Cash family's story begins in London where Grace and her siblings are born. In the early 1880s, they move to Thornton Heath, near Croydon, where they find other religious Nonconformists and social activists.

You might say Grace was lucky. She was born in 1872, just as a wave of social change was about to sweep the country. If she had begun life a few decades earlier, she would have known all the constraints of middle-class Victorian womanhood; the social, political and cultural limitations of a life lived predominantly at home, in the shadow of fathers and brothers. As it was, she reached adulthood during the Victorian *fin de siècle*,[1] a time of massive social upheaval, when the New Woman[2] took to her bicycle, middle-class girls could attend affordable day schools for the first time, and the railway transformed working lives forever.

She had the good fortune also to be born into a family where personal aspirations were high. Her parents, James and Elizabeth Cash, both came from the East End of London. James was the son of a tradesman (his father was a stay maker) and after their marriage, he and Elizabeth settled in Lower Clapton, in Hackney. The family's three-storey terraced house still stands in Cricketfield Road, a short walk from Hackney Downs. Cricket was popular and their road marks the site of an early pitch. It was in this house that James and Elizabeth's four children were born. Jessie Elizabeth arrived in 1869, Kate Gertrude the following year, then Grace Marion, and finally, Henry James in 1874. The family's prospects were good. James was employed as a commercial clerk, in a white-collared office job that positioned him firmly within the ranks of the lower middle classes. It was a significant step up from his father's world of the skilled artisan.

The lower middle classes sought upward mobility; they were status-conscious and they were strivers.[3] In Hackney, the Cash family were poised to take advantage of social and economic changes that would bring them acceptability and

Illustration 1.1   Elizabeth Cash with the children in about 1876, beneath chalk cliffs on the south coast of England. Clockwise from bottom left, they are Henry (with the ball), Kate, Jessie and Grace

underline their separation from the working class. Some forty or fifty years earlier Lower Clapton was regarded as a pleasant place to visit or even retire to. The Downs had been open country, with ploughed fields and crops of wheat. Now, in the 1870s, it was in transition and the semi-rural village was turning into a London suburb. The railway was chipping and tunnelling its way through large swathes of nearby countryside, continuously demolishing and disrupting. A network of new lines and suburban stations allowed those who could afford it to move out to the neighbouring counties, and commute back to the city each day for work. Often the lower middle classes followed the example of wealthier middle-class people by removing themselves further from the centre of towns and cities. Poorer residents were left behind so that by the 1880s Clapton had lost its appeal for the better off and ceased to be fashionable. More and more people were leaving for rural districts and eventually the Cash family would themselves join the exodus.

In the meantime there was the children's education to consider. Unlike many of his contemporaries, James Cash believed in educating his daughters properly. The school he and Elizabeth chose for the girls was one of the first set up by the Girls' Public Day School Company (GPDSC).[4] It included a kindergarten and was in Lower Clapton Road, a short walk from home. The three little Cash sisters would have made their way to the end of Cricketfield Road each morning, and then turned left beside the large Presbyterian Chapel (which later became the New Testament Church of God). Probably they were accompanied by one of the servants, or their great-aunt Mary (who was living with them). From the Presbyterian Chapel the most direct route was along Downs Park Road, across Clarence Road and then down Goulton Road. Lower Clapton Road itself was a main route for traffic in the 1870s, and the children had to negotiate it with care. Horse-drawn trams, at times carrying as many as fifty passengers, regularly passed with a clatter, on their way from Mile End Gate to Stamford Hill. An examiner at Grace's school observed at this time that the noise of passing tramcars was 'a most decided element of evil'.[5]

At 179 Lower Clapton Road, not far from the school, was one of Hackney's most striking Georgian houses, built by a pottery manufacturer who had made his fortune from chamber pots. Two suggestively shaped flower pots, one each side of the gateway, confirmed his calling and unimpressed locals referred to the house as Pisspot Hall. By the time Grace and her sisters were at school nearby it had been taken over by a charity and is marked on maps of the time as 'The British Asylum for Deaf and Dumb Females'. Victorian society kept its casualties well-hidden and it is unlikely that the children ever saw any of the unfortunate inmates.[6]

By the end of the nineteenth century the school in Lower Clapton Road had also closed. No records survive, but we can suppose that the drift of the middle classes to suburbs further afield was chiefly responsible for its demise. The head teacher, a Miss M. Pearse, remained in post until 1892, but by then the school had lost many of its target population, amongst them the Cash family.

It was the early 1880s, and Grace was about nine years old when James and Elizabeth took their children away from Hackney to live in the village of Thornton Heath, set in attractive countryside in the municipal ward of Upper Norwood, near the town of Croydon. Croydon was well served by the new railways and its population was growing quickly. Thornton Heath, too, was in transition and would soon be transformed from a village into a suburb. Most of the incomers to the district were prosperous merchants, clerks (like James) or tradesmen but there were also domestic servants and agricultural workers who drifted in from outlying rural areas. Life could be lived more healthily here – or so the middle classes believed. With its gardens and open spaces, suburbia was promoted as a rural idyll. In Upper Norwood, 'the fresh air suburb', as the poster in Illustration 1.2 describes it, the Cash family would have hoped for a more leisurely pace of life, a retreat from the dirt, bustle and noise of the city.

As well as its supposedly fresh air, Norwood was known for its villas (or large suburban houses), most of which had been constructed in the 1850s and 1860s, not always with the approval of local inhabitants. The suburb is the setting for Arthur Conan Doyle's novel, *Beyond the City*. In this 1890s story, two middle-class Victorian sisters, obliged to sell their family estate, watch 'with sore hearts' as three villas are built on their last remaining field. Gradually,

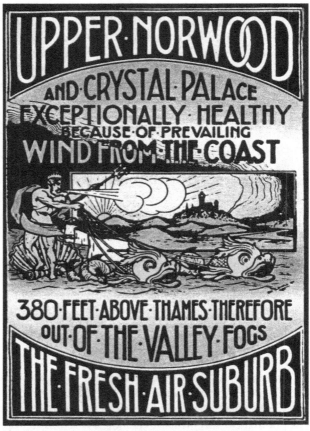

ADVERTISING POSTER 1900 © ERIC PRICE

Illustration 1.2   Poster advertising the 'fresh air' suburb of Upper Norwood in about 1900

the reader is told, 'the City had thrown out a long brick-feeler here and there, curving, extending, and coalescing, until at last the little cottages had been gripped round by these red tentacles, and had been absorbed to make room for the modern villa'.[7]

The Cash family could not afford a villa but their house in Decimus Burton Road was only a few hundred yards away from Grangewood Park, with its woodlands, gardens and muddy tracks. Just as in Hackney, here they lived on the edge of the natural landscape, which the children were free to explore. Perhaps Decimus Burton Road (named after a successful young architect, the tenth child in his family) sounded to them like a grand address. The naming of Victorian streets could provide a subtle clue to the status of a particular locality and most in the lower middle classes were intent on establishing their credentials in the suburbs, sometimes embellishing and decorating their houses in ways that were meant to mark the social standing of the inhabitants. Certainly, the Cash family now had more space, but in reality the semi-detached house, with about five rooms, was unexceptional. Next door was the water board's pumping station (steam-driven and undoubtedly noisy) and a local reservoir.[8]

In the late nineteenth century, people might find themselves in Thornton Heath on their way up the social escalator, as the Cash family did, but they might also find themselves there on their way down. When, in 1892, Helen Corke and her middle-class parents came to live in Clifton Road (just two streets away from the Cash family), they were struggling with adversity. Her father's grocery business had recently failed and the family had been forced to relocate from the country so that he might find employment. Corke remembered that her mother particularly resented the drop in social status:

> My mother's home talk is a daily recital of privation. One day a small black kitten comes in, begging for food. I feed it but with a burst of anger my mother drives it out – she cannot afford, she says, to feed stray cats. The day following I find the kitten dead in the gutter. I go cold with hate against my mother, as if she were the visible symbol of our poverty.[9]

Corke had literary aspirations and would later become known for her close friendship with the writer, D. H. Lawrence.[10] And she did eventually become a writer herself, but in the late 1890s she had few employment options and reluctantly took up a teaching career.

Unlike the Corkes, the Cash family had every reason to feel optimistic. James held secure employment with a commercial stationery company in the City throughout the 1880s. Each morning he joined the lines of businessmen on the platform at Thornton Heath station and took a commuter train into London. Like other middle-class suburban children, Grace and her siblings grew up expecting their father to be absent all day. Elizabeth was left in Thornton Heath to run the home (with the help of a servant) and to see to the children. James would reappear in the evenings when he was able to retreat from London and workplace demands.

Clerical work was a popular choice for many young men. Opportunities in London were increasing rapidly with the growth of government, the expansion of big business and the globalisation of trade. James was a loyal employee and his length of service enhanced his promotion prospects. Average salaries for clerks were rising and in the 1880s James was probably earning about £150 p.a., with yearly increments.[11] In the second half of the nineteenth century this figure was less than the bare minimum needed to maintain a middle-class way of life (including servants, fee-paying education for the children, appropriate housing, dress, furniture and entertainment), and it placed the Cash family on the lower echelons of the middle-class ladder. Elizabeth and the children relied on James for financial support and though the family later remembered that she was always short of money, Elizabeth would probably not have considered working for wages outside the home, for fear of undermining her husband's status as provider.

So the Victorian social world was split, starkly divided into domestic and public, female and male. Grace would have seen her father briefly in the evenings, but otherwise only at weekends. The writer Katharine Chorley grew up in a prosperous suburb of Manchester and remembers that after the 9.18 train had left the local station her neighbourhood became exclusively female: 'You never saw a man on the hill roads unless it were the doctor or the plumber, and you never saw a man in anyone's home except the gardener or the coachman.'[12]

By 1892 James was company secretary to Millington and Sons Limited, a well-established business for wholesale and export stationery.[13] It was an impressive achievement. From his East End beginnings, to office employment as a junior clerk, he now held a senior post in a successful private firm in the city. He was amongst the first subscribers of Millington's and remained a shareholder for most of his life. The Company grew rapidly as wholesale and export orders for quality stationery increased; in 1897 there were sixteen shareholders, and by 1902 there were twenty-five. By 1908 Millington's had several warehouses and factories in various parts of London and a warehouse in Birmingham. In 1911, at a meeting of some of the Company's directors at a country house called Basildon Park, in Berkshire, a decision was made to introduce a new quality writing paper and 'Basildon Bond' (as it is still known) was launched. By now James Cash had retired from the Company but he was still a shareholder and benefitting from their success.[14]

Stationery was important to the Victorians. Its use underpinned contemporary codes of polite behaviour and marked the social place of individuals and families. Calling cards, printed invitations and writing paper were all in fashion and stationery manufacturers themselves might give guidance on such things as the etiquette of letter-writing. In *The Diary of a Nobody* (which first appeared as a comic serial in *Punch* magazine in the late 1880s), Charles Pooter (a London clerk) is mercilessly mocked as he betrays both his ignorance of the social rules and his lower-middle-class ambitions. When, to his wife's tearful pride, he receives a printed invitation from the Lord Mayor and Lady Mayoress to a ball at Mansion House, he cannot contain himself. First, he must seek advice about the appropriate way to answer the invitation. Other preparations include buying a pair of lavender kid gloves 'and two white ties, in case one got spoiled in the tying'.[15] His wife, Carrie, sends the invitation for her mother to look at, only to have it returned stained with port. And the ball itself is a humiliation. Pooter discovers his local ironmonger has also been invited and – worse still – that the man is on good terms with some of the more distinguished guests. Pooter himself knows no one and after sampling 'any amount of champagne' he slips and collapses on the dance floor, pulling Carrie down with him.[16]

James and Elizabeth, and their friends, may well have been familiar with *The Diary of a Nobody*, which helped to establish a genre of popular fiction based on mockery of the lower social orders. Of course, the self-important Pooter, with his clumsy pursuit of status and respectability, was a stereotype and the lower middle classes, so long parodied and pilloried by social commentators, in reality were a diverse group. Most cared desperately about respectability, however, and at times found themselves trapped by social conventions which they could not question.

In a short story called *A Daughter of the Lodge* (1901), the writer George Gissing creates a lower-middle-class character that is not comic like Pooter, but resentful and insecure. After two years away, May Rockett, the able and ambitious daughter of the head gardener at Brent Hall, returns to visit her family who have lived for many years in the Lodge on the estate. Her parents are keen that she should behave 'in a suitable manner' towards 'her feudal superiors'. But anger at a perceived slight from Hilda, the daughter of Sir Edwin and Lady Shale at the Hall, prompts a hostile reaction from May as she returns home one evening:

Just as her hand was on the gate a bicycle-bell trilled vigorously behind her, and, from a distance of twenty yards, a voice cried imperatively – 'Open the gate, please!'

Miss Rocket looked round, and saw Hilda Shale slowly wheeling forward, in expectation that way would be made for her. Deliberately, May passed through the side entrance and let the little gate fall to.

Miss Shale dismounted, admitted herself and spoke to May (now at the Lodge door) with angry emphasis. 'Didn't you hear me ask you to open?'

'I couldn't imagine you were speaking to *me*', answered Miss Rocket with a brisk dignity. 'I supposed some servant of yours was in the vicinity.'[17]

The challenge does not go unpunished. May's parents are immediately told the family will be evicted from the Lodge and they are saved only by their daughter's humiliating climb-down and abject apology. In anguish, May is forced to acknowledge her inferior status and swallow her pride. The lower middle classes lacked the financial means to circumvent Victorian social codes and Gissing's story underlines their insecurity. An acceptance of their social position and a clear recognition of their dependence on the better off were necessary if the lower middle classes were to avoid disaster.

The Rocketts' wish that their daughter should behave 'in a suitable manner' towards the family's superiors highlights the Victorian preoccupation with respectability. A 'respectable' life was characterised by independence, hard work and cleanliness, the avoidance of alcohol, gambling and swearing, and by regular attendance at church or chapel.[18] 'Respectability' was not fixed; it was not solely an economic classification and could include those at either end of the social hierarchy, not just in the middle. To the 'respectables', as they were sometimes called, observable behaviour mattered. Within the Victorian lower middle classes, as Gissing suggests, an insistence on your 'respectable' status was one way to counter a sense of threat. You might not have the trappings of wealth and privilege, but you could treat others properly and insist on being treated properly yourself. Essentially, it was a way of living with inequality.

Some, like May Rockett, understood the unfairness that was at the root of the English class system and rejected 'respectable' values, seeking instead to define respectability in their own terms. In a counter-attack, some turned to socialism and became innovators. By 1890 religion was becoming less and less a necessary part of 'respectability' and as the lower middle classes, who did not have as much to lose as the more elite, drifted out towards the suburbs, their family ties were loosened and they found like-minded friends. So it was with the Cash family. Like most lower-middle-class church-goers at the time, they were religious Nonconformists;[19] they believed in religious equality and in churches that were free from outside influence. Members of such groups were especially likely to hold radical political views. Now, pocketed in a corner of Thornton Heath near Croydon (and later in Coulsdon) the Cash family associated with other Nonconformists and with activists, some of whom were pushing for the abolition of poverty, others for a less formal family life, and some who were earnestly searching for the key to human existence.

It was an exciting time and it is difficult now, with our modern sensibilities, to imagine the passion and energy that went into these activities. Croydon is now a large, diverse town with city aspirations but in the 1880s and 1890s it was home to some quite distinctive political communities. The socialist movement was endlessly organising and reorganising itself, busily inventing new groups, societies and movements, particularly religious ones, and Croydon was ripe. The town was expanding and many of the newcomers, like the Cash family themselves, were from parts of nearby London which had relatively low attendance at the orthodox churches and high levels of interest in the activities of smaller religious groups.

In these circles marriage might be more freely contracted than in other classes and it might be more companionate. Girls were less discriminated against. Typically, Nonconformists mixed with their friends and associates at the local chapel and both Elizabeth and Jessie appear briefly in the records of the Congregational Church in Thornton Heath,[20] where the Reverend William Jupp was the minister. It was a short walk from their home. Jessie was a conscientious helper in the Sunday school, and in the late 1880s she organised a church bazaar and a Literary Society. Her local church, like many other congregational churches in the Victorian suburbs, was alive and buzzing. At the church bazaar, visitors might join in hymns and prayers, but also enter competitions (in 'Ladies' Pencil Sharpening' or 'Gentlemen's Needle Threading'). Or they might go and see some 'electrical novelties', including many 'startling' new appliances. Alternatively, they might sit quietly and enjoy 'pleasant music' at an evening gramophone concert.[21]

Many in the congregational churches had a zeal for self-improvement and were in search of a more fulfilling life. In 1889 Jupp enthusiastically endorsed Jessie's application to Stockwell College in south London where she would soon go to train as a teacher. And Elizabeth Cash, though her own circumstances were constrained, was busy at the church too, with the Dorcas Society, a charity that collected clothing and then sold it to the poor at reduced costs. Here, in the early 1890s, she worked alongside a Mrs Oakeshott whose husband had recently retired from his job as Postmaster in Sunderland. Eliza and Joseph Oakeshott had recently moved south to Surrey to join their eldest son, Joseph Francis, and daughter, Violet, in Thornton Heath. With them had come their youngest son, Harold, aged about twenty.

In a small community most paths were bound to cross sooner or later. A short walk from the Cash family home in Decimus Burton Road was Beulah Road, and number 40 (known as 'Oak Villa') was by now the centre of operations for a new organisation that had recently relocated from London, called 'The Fellowship of the New Life'.[22]

The Reverend William Jupp was a member of the Fellowship, and may well have spoken about it to his parishioners, perhaps encouraging them to attend meetings. Jupp was by now struggling with the constraints of formal church membership and the aspirations and ethos of the Fellowship caught his interest.[23] James Cash had also found his way to Oak Villa. Here, he and his family now regularly associated with the Oakeshotts (amongst others). Joseph Francis Oakeshott and Maurice Adams (an insurance agent with a passionate interest in moral philosophy) would prove to be the Fellowship's mainstays.

Amongst the lower middle classes, voluntary organisations and societies were now flourishing. The Fellowship of the New Life had both spiritual and political aims but other groups appealed to more secular interests. The Literary Society that Jessie ran, for instance, would have enabled some to remedy deficiencies in their state-funded education. The evening gramophone concerts would have helped make good their sense of cultural inferiority. As wages rose in the last quarter of the century, the middle classes found they had more opportunities for recreation, even if it was somewhat earnestly pursued. Visits to the seaside had health benefits and travel to the continent could be intellectually stimulating. The photograph in Illustration 1.1 suggests the Cash family's visit to a beach was something of an occasion and, that like most of their Victorian peers, even in this remote location they are intent on preserving decorum and privacy.

None could have known what lay ahead for these serious-minded and aspirational people but in Grace's early world, a value was placed on companionship, respectability and the urgent need for social change. Peter Bailey, a social and cultural historian, comments that 'of course, a suburban nonconformist is not the same as a social rebel',[24] but the emphasis in such circles on fellowship, belonging, camaraderie, freedom, equality and self-improvement indicates at least a readiness to push against established boundaries.

2
Learning Curves

Grace and her siblings go to progressive, fee-paying schools, and there is an early glimpse of Harold, when he is still living in County Durham with his parents. He is at school in Newcastle.

The Cash family's move from Hackney to Thornton Heath in the early 1880s may have been prompted partly by a desire for a healthier life (like the one promoted in the poster in Illustration 1.2) but probably uppermost in their minds was the need to find good secondary schools. Like their peers, James and Elizabeth knew that formal education was (as it is still) the chief marker of social status, and the best way to ensure their children's progression. The English middle classes were growing in number and gaining power and wealth. As they spread out to suburbs like Thornton Heath their standard of living improved and their demands grew. There were too few affordable secondary schools to meet their needs and those that did exist were unevenly distributed geographically. There was no coherent 'system' and no central authority to oversee quality or standardise the curriculum.[1] Still, change was underway and if we pause here to trace some contemporary debates, the Cash family's educational experiences, and those of other families like them, can be seen against a wider backdrop.

Some years before Grace and her siblings arrived in Thornton Heath, Matthew Arnold (who is best remembered as a poet and writer but who was also an inspector of schools and a trenchant critic of England's educational institutions) had urged the nation to organise its secondary education. The English middle classes were, Arnold believed, almost the worst-educated in the world, their schools inadequate and unsatisfactory. He had studied schools on the continent and found them to be far superior. Competition from abroad was increasing and Arnold was open, too, about his fear of social unrest. He believed educational reform was needed to see off the threat of revolution. If England was to 'live and prosper' then immediate action was required. 'Organise your secondary

instruction', Arnold exhorted his countrymen.[2] He would die before any meaningful organisation was attempted but his investigations contributed to an
influential report on England's schools. In 1868 the Taunton Report (as it was
known) acknowledged the inadequacy of contemporary middle-class education
and over the next two decades or so, changes began that would transform the
life chances for hundreds of young people like Grace and her siblings.

One of the report's findings was that 'endowed schools' (those which were
maintained by funds from charitable foundations, such as hospitals) varied
greatly in their quality and efficiency. Many of these institutions were no
longer addressing their founders' aims; others were finding their endowments inadequate; some had been forced to close. The Taunton report proposed restructuring the endowments of these schools to meet modern needs;
eventually, new schools emerged and many others were reorganised and
invigorated.

Two of the young people in this story would soon benefit directly from
these reforms. In 1877, following the union of four foundation schools in
central London, a new institution (known initially as the United Westminster
(Endowed) Schools)[3] was opened and in 1884, at the age of ten, Henry Cash
(Grace's young brother) became a pupil there. By the end of that year there
were 760 boys in attendance, all aged between seven and fifteen.[4] The school
was fee-paying and although entrance to the four original schools had been
restricted to children brought up according to the doctrines of the Church
of England, now admission was open to those from other faiths, including
Nonconformists like Henry, on the basis of merit alone.

By the time Henry arrived, as well as the usual secular subjects such as history, composition and mathematics, the school had a strong focus on industrial and technical training. It was building on the vocational work carried
out in one its constituent schools (St Margaret's), and now its novel teaching
methods and distinctive curriculum were attracting considerable interest. The
school was one of the first in the country to teach science in laboratories and
by the late 1890s it was advertising its 'well-fitted boys' workshops' where
carpentry, turnery and metal work were being taught by expert craftsmen.[5]
Robert Goffin, the school's headmaster, was liked and respected. He had himself studied under some eminent Victorian scientists (including the biologist
Thomas Huxley, who was a strong supporter of Charles Darwin's work on
evolution).

Along with Goffin's commitment to science went his belief in the importance of a sound general education as the basis for any future career, whether
commercial, scientific or legal. In the 1880s, when they left the school, some
boys took up places at London University; some joined the civil service. Others
moved on to technical colleges, and Henry Cash was amongst these. In 1889

he enrolled at Finsbury Technical College in London, and three years later he emerged as a fully qualified electrical engineer.[6]

In the meantime, at the other end of the country, a young Harold Oakeshott (Grace's future husband) was also studying at an endowed school. The Royal Grammar School Newcastle (RGS) had been founded during the reign of Henry VIII with the help of charitable funds from the local St Mary's Hospital. As in most English towns at this time, the hospital was the centre of charitable activity and over many years, the fortunes of these two institutions were to remain inextricably linked.

Following the Taunton Report the 1870s might have been a decade of progress for RGS, and the school might have found its feet as the endowed schools in Westminster did. There was now a national focus on secondary schooling and intense debate about the principles of education. RGS had just moved to newly built premises in a district known as Rye Hill and the local Council was keen to improve the school's standing. Its academic work needed strengthening and the Council also wanted the school to become a 'social ladder', to offer the kind of education that would enable poorer students to enter the professions.

There was some limited progress in the 1870s towards this goal (and a few more students entered higher education) but it was a difficult decade. Complex legal wrangling left the school without any endowment at all, and they were entirely dependent on pupil fees which poorer families could ill afford. The endowment payments were eventually restored by a High Court order in 1881 and when Harold Oakeshott came to live in nearby Sunderland two years later, with his parents Joseph and Eliza, a new headmaster had just been appointed. Samuel Logan, a Cambridge graduate aged 38, was the first headmaster not to be an ordained member of the Church of England. Until his arrival, Nonconformists (like the Oakeshotts) had been reluctant to send their sons to the school. Now there was a new man in charge, modern buildings, better qualified staff and every reason to hope that adequate funds would soon flow. It is easy to see why Joseph and Eliza would have considered the school a good prospect for their youngest son.

The family had come to the north-east of England because Joseph was to take up a new position as Sunderland's Postmaster. He had already held similar posts in London and this would be his last before retirement. Harold began his studies at RGS in 1884, aged about 13, and though details of his activities are scant, he was clearly an able pupil. A slight speech impediment (he could not pronounce his r's properly) did not hinder his academic progress. His thick, wiry curls receded quickly once he reached adulthood and in later life he was seldom seen without a beret which he wore to cover his baldness. He spoke with what his family many years later described as a 'posh' accent.

Illustration 2.1   Harold Oakeshott as a young man

The boys at RGS followed a traditional academic curriculum, including Latin and Greek, and Harold developed scholarly interests.[7] He passed his Cambridge locals and won two prizes, one for French and the other for proficiency in general subjects. The books he was awarded (including a recently published edition of *Lord Macaulay's Essays and Lays of Ancient Rome*) stayed on his bookshelf until the end of his life, as part of a collection that reflected a keen interest in history and politics. It included many texts that became and would remain classics, such as John Stuart Mill's *Considerations on Representative Government* and *Past and Present* by Thomas Carlyle.

Harold had grown up in a house filled with books. In the 1880s, as RGS continued to struggle, the call went out to pupils to bring from home whatever books they could spare. Several times, Harold donated parcels of books to the school's grateful librarian. By 1887, however, there were still just 85 volumes in the library – and about 250 boys enrolled. Logan was a progressive headmaster and keen to see the school's facilities improved. He set up a school magazine (*The Novocastrian*) and promoted extra-curricular activities, like drama and sport, but he was always constrained by lack of funds. Despite the High Court ruling in the school's favour and the restoration of the endowment payments, RGS still had insufficient money for its needs and although it was attracting better qualified staff, it could not retain them. The endowments from the hospital, even when paid regularly, were not enough to cover the salary bill. It was not until 1888 (a year after Harold had left the school) that agreement was reached and in a landmark ruling it was decided that the hospital's charitable funds would be devoted principally to the school. As a consequence its endowment income was hugely increased.

Joseph and Eliza would not have expected local interests to determine the school's fortunes for as long as they did and they would not have expected it to be held back in the way it was, mired in a soft bog of entrenched rivalries and historical agreements, some dating back centuries. The Taunton Report had seemed to pave the way for families like theirs and although it did create new opportunities for many boys like Harold in many parts of the country, in other places it took much longer to see the improvements the middle classes were now demanding. In London, Henry Cash had been luckier.

Despite the lack of funding and facilities at RGS, however, Harold was successful enough. He seems to have played a limited part in school activities and showed some small measure of enthusiasm on the sports field (when in 1885 he was unplaced in the sack race). His parents could not afford to send him (or his siblings) to university and when he left RGS, the school expected him instead to make a career in the civil service. In fact, by 1891, he was employed as a tea merchant's apprentice.

But his education was by no means over. While Harold was still at school, his father had become active in the local university extension movement. In the late nineteenth century Cambridge University became the first institution to offer lectures to diverse groups of adult learners in provincial centres like Sunderland. After a full day's work at the post office, Joseph often spent his evenings at the library hall, introducing visiting academics, chairing lectures and leading discussions. Thanks in large part to Harold's father, and others like him, if you lived in Sunderland in the 1880s and 1890s, you could sign up for Cambridge University courses (for example, on the history of France and the Netherlands during the Reformation) and in 1888, Harold's sister, Violet, received a certificate for successfully completing a course in 'ancient comedy'.[8]

It is likely that Harold, too, took some university extension courses once he had left RGS. The movement was burgeoning during the 1880s and although the courses had initially been devised for adults, they were also attended by schoolchildren, especially girls, and young school-leavers who were not yet settled in their careers. Clerks and shop workers took up the chance to study in their evenings – and in Newcastle some people travelled as far as ten miles to hear the lectures. The courses were especially popular in the north of England

and miners in Northumberland were amongst the most responsive learners. One pitman wrote about his experience:

> I deeply deplore the last thirty-five years of my life. Being buried in the mines since I was nine years of age, and taught to look jealously on science as being antagonistic to religion, I little thought what pleasures of thought and contemplation I had lost; I have however broken loose from my fetters and am proceeding onwards.[9]

The working classes were hungry for knowledge. Now at last, wrote Oscar Browning in 1887, the old English universities, too often considered 'hot-beds of clericalism and toryism'[10] could connect with the nation's people. Higher education would no longer be the preserve of the privileged. Undoubtedly, the Nonconformist Oakeshott family endorsed these sentiments and Harold, like his father, was already an activist. One summer, the local newspaper carried a report of a Fabian Society picnic,[11] involving a boat ride on the river and what was referred to as a day of a 'realised socialism'. The outing was an 'unqualified success', due largely to the efforts of Harold and two of his comrades.[12]

By now, Joseph's health was beginning to fail and following his retirement in 1891, the family moved back to the south of England. In Thornton Heath they settled near Joseph Francis and Violet (who had travelled from Sunderland some months earlier). Harold, too, took up residence locally and quickly established links with several socialist organisations, notably 'The Fellowship of the New Life', in which his eldest brother was by now prominent. His father died just two years later. Joseph had been admired for his 'genial courtesy' and known for his effective management of the Sunderland post office.[13] He had had such cordial relations with his chief clerk that it was thought the two 'would easily pass for father and son'.[14] Now, the news of his death came as 'a painful surprise' to friends and associates back in the north-east.[15]

It was not just boys who benefitted from the late nineteenth-century tide of educational reform. While Harold was at school in Newcastle, Grace and her two sisters, Jessie and Kate, were also embarking on their secondary education. The 1851 Census had revealed that there were half a million more women than men in England. It followed that not all the young women would find husbands and if their daughters were not going to marry, then middle-class fathers needed them to be educated so that they could support themselves. Hardly any lower-middle-class families were sufficiently well off to keep their daughters at home for any length of time, and the Cash family was no exception. James would have wanted his daughters to go to secondary school for practical as well as liberal and egalitarian reasons.

The Taunton Report provided official evidence that change was required. As well as discovering that many endowed schools were struggling and in need of reorganisation, Taunton also concluded that secondary education for girls was especially limited. The Report proved a turning point and its finding that there were only thirteen girls' secondary schools in the country gave campaigners further grounds to push for improvements.

And improvements were badly needed. In the words of Frances Power Cobbe, the Irish writer and feminist campaigner, education for women in the middle of the nineteenth century was 'more shallow and senseless than can easily be believed'.[16] The scholarly activities of middle-class girls just a generation older than Grace and her sisters had been very restricted. Before they reached the age of ten, they would have received some instruction at home, perhaps from a nurse. After that age, they would probably have gone to a small local day school, perhaps taking up to twenty pupils and run by middle-class women in a private home. The curriculum (if we can call it that) would have been modelled on middle-class family life. Young girls, like Jessie, Kate and Grace, would have learnt little more than reading and writing, with some needlework and drawing. Now, in the wake of the Taunton Report, came new impetus for reform and at a public meeting in the Royal Albert Hall, the Girls' Public Day School Company (GPDSC) was founded. It was 1872, the year Grace was born, and at last there were to be affordable day schools for girls. She and her sisters would have a proper education.

The Company's policy was to go only where a school was asked for and where residents had already demonstrated their good faith by forming a committee and taking up enough shares to cover initial expenses. The Cash family was still living in Hackney at this time and, in 1875, when the new school in Lower Clapton Road was proposed, James Cash bought a share in the Company. There would have been little doubt in his mind about the need to educate Henry but middle-class girls had so far lacked the opportunities enjoyed by their brothers – and James now invested in his daughters' futures. The fees were set as low as was compatible with the school being self-supporting. As an investor, James could sell his share again (and he did so in 1903) but he would not have expected to profit financially.

The new Company had a motto for its young scholars: 'Girls, knowledge is now no more a fountain sealed. Drink deep!'[17] They planned an academic curriculum (to match what was on offer in boys' schools) and decided there would be no discrimination in terms of religious affiliation or social class. In reality, of course, the venture principally attracted the middle and lower middle classes since they could afford the fees. The upper and upper middle classes could afford to employ governesses or send their daughters to fashionable boarding schools, followed perhaps by finishing school in Europe. And, although eventually limited financial support for working-class pupils did become available, generally such parents could not afford school fees at all; it was well into the twentieth century before large numbers of working-class girls had access to academic secondary education.

Years later, one of the first Company school pupils remarked: 'It is not easy to realise how great was the innovation that we should be thus all assembled together in a public school.'[18] Although only a small proportion of middle- and lower-middle-class girls were lucky enough to benefit from the initiative, it was a successful model for setting up schools and it broke new ground. Other public bodies quickly adopted it and the 'high school movement' took hold. New schools proliferated.

After the Cash family's move to Thornton Heath, the obvious choice for the girls was another GPDSC school, the local Croydon High School for Girls (CHSG). It was the third Company school to open (the first outside London) and it had begun in 1874 with just 88 pupils. When the Cash family arrived in the district, it had recently moved to new premises at 36 Wellesley Road. There were now 230 pupils on the roll and the school was still growing.[19] Jessie, Kate and Grace all enrolled in March 1881.[20] The girls had a straightforward journey to the school, thanks to the recent arrival in Croydon of horse-drawn trams. The first section of tram line from Thornton Heath Pond to North End (in central Croydon) was already in place. Soon, a branch line opened that took you from the town centre out along Thornton Heath High Street, and it was a short walk from there to the family home in Decimus Burton Road.

The school's headmistress was Dorinda Neligan, the Irish-born daughter of an army lieutenant. When she took up the post, she was just over forty years old. Like most of her generation, she had had little opportunity to acquire a formal education herself and at first she felt insecure. Years later, she remembered feeling that she had spoken badly at the school's opening: 'I was too *self-conscious*', she wrote to a friend, 'too burdened with my own personality – a fault I hope I have somewhat overcome now.'[21] In fact, Neligan was ideally suited to her role. She had a commanding presence, a fine voice, keen blue eyes and a strong sense of justice. During the Franco-Prussian war from 1870 to 1871, she had worked for the Red Cross as a nurse. On one occasion, in the cold basement of a school that had been adapted to serve as a temporary hospital in Metz, she defied orders by burning the desks to provide warmth for the wounded soldiers. Another time, she had promised to be at a soldier's bedside when he died but she was off duty when it happened. So that evening she crept outside to the shed where the corpses were laid, found the boy and cut off one of his coat buttons. She still had the button when she was headmistress at Croydon.[22]

In school, Neligan usually wore a black silk dress and often a lace cap (or mantilla). Under her leadership, the school had few rewards or punishments. She believed that the girls' love of learning alone would motivate them. In 1885 she explained to one visitor that the system worked and that the girls behaved well: 'After all, nobody *wants* to forget books or to be untidy, and really there's no fun in being unruly if you're not punished for it, is there?'[23]

She had her critics, however, including some of the shareholders, who believed she was not enough of a disciplinarian. They did not want to see strong-minded women encouraging the girls to be dissatisfied with life at home. Neligan remained a force to be reckoned with, however, and she overcame her detractors. She knew all her pupils by name and she actively worked to transmit her personal values. After her retirement from the school, she joined the women's suffrage movement and as a member of the Women's Tax Resistance League vigorously opposed the taxation of those who could not vote.[24] Her passion for education, her care for the less fortunate, her belief in justice, her courage and independence were all remembered by those who mourned her death in 1914, just before the outbreak of the First World War. A former pupil remarked: 'We had a wholesome fear of Miss Neligan and never dared do anything of which she disapproved – and we loved her with all our heart.'[25]

What sort of pupil was the young Grace? When she first arrived at the school, she had just turned nine and she stayed until she was nineteen. During her ten years under Neligan's regime she flourished. The Company undertook to prepare girls for public examinations, like those undertaken by their brothers. By this time, at Croydon, German could be taken as an alternative to Latin, and Chemistry (previously considered too dangerous for girls) was introduced a short time later in 1885. In 1887, at the age of fifteen, Grace passed Cambridge Local examinations in English, French, German, Euclid (now called Geometry), Algebra and Drawing. In the next couple of years, she also passed government examinations in Art and Chemistry.

She was a confident all-rounder and prominent, too, in school prize-giving events and entertainments. On these occasions pupils of all ages performed a variety of pieces, including violin and piano solos, scenes from Molière or choruses from Mendelssohn. Twice, Grace recited passages from Goethe and in 1886, she gave a recitation as 'Portia', from Act IV, Scene 1 of Shakespeare's *Merchant of Venice*. Shakespeare's heroine is wealthy, spirited and intelligent. Whilst disguised as a man, she makes a famous speech in this scene about the need for compassion in judging others ('The quality of mercy is not strained') and her eloquence and ingenuity win the day. Grace may well have found the role of Portia attractive. Any schoolgirl might.

Illustration 2.2 Senior pupils at Croydon High School for Girls, *c*.1888. Grace Cash is fourth from the left in the back row and the headmistress, Dorinda Neligan, is seated in the centre (with cat)

During Grace's time at Croydon High School for Girls, as is evident from the photographs, there was no school uniform. Grace had long, auburn hair (her nickname was 'copper top') and the one school requirement was that hair must be tied firmly back or piled on top of the head. Anyone forgetting her hair ribbon was sent to Miss Neligan and supplied with string.[26] The girls dressed according to fashion and their resources. Puffed sleeves, sashes and bows were popular. The Rational Dress Society was actively campaigning throughout the eighties for a style of clothing that did not impede women's movement or injure their health and they opposed tight corsets, heavy dresses and high heels. The students in the photographs here would have been aware of the controversy. In their full-length dresses, high narrow collars and fitted bodices, they are not frivolous or radical. Their clothes, the books they clasp, the expressions on their faces – all underline their seriousness. They are presented to us as scholarly and determined young women.

Illustration 2.3 The Upper Sixth Form at Croydon High School for Girls in 1891. Grace is standing in the back row, on the right, arms behind her back

Unusually for the time no religion was taught, which concerned some parents but met with approval elsewhere, for example, within the Jewish community. As Nonconformists, James and Elizabeth Cash would also have welcomed the lack of formal religious observance. There was an insistence on classical subjects, such as Literature, Latin and Mathematics, but unlike their brothers, the girls had also to acquire artistic 'accomplishments'. The academic, Ellen Jordan, argues that middle-class girls were taught drawing, art and music in an effort to imitate the education of aristocratic women. The high schools were expected to turn out pupils who would enhance their family's standing while they lived at home and eventually attract husbands of the same social rank.[27] They offered little challenge to the assumption that women needed to be educated principally so they could be good wives and mothers, and to safeguard the influences of family life the compulsory academic work took place only in the mornings. It was well into the twentieth century before the Company introduced afternoon school. But with the arrival of the new high schools, life chances had dramatically improved for some lucky middle-class girls, and many, like Grace and both her sisters, worked in later life to extend the benefits of a full education to others less fortunate.

By the time the photographs in Illustrations 2.2 and 2.3 were taken, both of Grace's sisters had already left the school. In 1883, aged about thirteen, Kate was first to go. She had been there for just two years and records suggest that she was suffering from the early symptoms of tuberculosis. Rather than keep her at home, so bringing her secondary education to a halt, James and Elizabeth managed to find the resources to send her to a private girls' boarding school in Ramsgate. The seaside town was described in a contemporary guide as having 'bracing air' and being 'naturally so well-drained' that it had become 'a favourite place for the medical profession to order convalescent patients'.[28]

Townley House Ladies Boarding School had been founded in about 1840 and it survived until after the First World War. The building itself was designed in the late eighteenth century by Mary Townley, artist and architect, and is still located in Chatham Street, Ramsgate. Private day schools and boarding schools were flourishing and by the end of the century, about 70 per cent of girls receiving secondary education were in these kinds of establishments, many more than were in schools run by the GPDSC.[29] The teaching could be variable and accommodation might be inadequate, but in the 1880s there were seven or eight such schools in Ramsgate alone.

Townley House appears to have been one of the more spacious; in 1904 it was advertised for sale with seven dormitories on the top floor, a drawing room, music room, study and classroom on the first floor, as well as a drawing room, dining room, large well-ventilated school room and two classrooms on the ground floor. There were tennis and croquet lawns. There is a story about a resident ghost – and a slightly more believable one about the infant Princess Victoria (later Queen) who used to visit. She is reputed to have dropped a full inkwell over the floor at the top of the staircase and then been made to scrub it clean. A brass plaque used to mark the spot.

The new school may well have suited Kate. She would turn out to be the most scholarly of the three Cash girls – but she was also the most vulnerable and at this time a move away from a day school with high academic expectations (and over 200 pupils) to the small Ramsgate boarding school (with about 20) gave her the chance to grow in a protective environment. The need for the family to find the resources to support Kate may have been behind the decision her older sister made a year later. Jessie was just fifteen in the summer of 1884 when she, too, left Croydon High School for Girls. In her study of families in late Victorian London, Dina Copelman observes that in the lower middle classes, parents and siblings often cooperated to ensure that all the children had the chance to establish themselves in worthwhile careers.[30] Jessie would have believed that, as the eldest daughter in a Victorian family, she had a responsibility to help her younger siblings. Her story resumes (as Kate's does) in Chapter 3 when they both set off on distinctive career paths of their own. Grace, meanwhile, stayed on at Croydon High School until 1891, and in the following year she entered Newnham College, Cambridge.

3
Finding Their Own Way

After leaving Croydon High School for Girls, the three Cash sisters are keen to pursue their studies further. They want to help educate others and each of them finds a distinctive way to do this, limited as they are by the family's resources and the constraints of late-Victorian womanhood.

James and Elizabeth Cash wanted all their daughters to have an occupation. They could support them for a time while they acquired skills and training, but their claims to respectability depended on the family maintaining a steady income. As the eldest, Jessie understood this. When she planned her career, she did not have a wide range of occupations to choose from. She might have become a nurse but at this time trainees had to be between 25 and 35 years old. Teaching would probably be satisfying but governesses were becoming obsolete with the growth of secondary education, and the new secondary schools, like the one she herself had attended, increasingly preferred their teachers to be graduates. Women with sufficient capital could start their own fee-paying schools, but Jessie was not in this position. Elementary schools (schools for working-class children) now provided opportunities for some pupils to progress to salaried teaching posts and in 1880, 79 per cent of these 'pupil teachers' were female.[1] Two years after leaving Croydon High School for Girls, Jessie resolved to join them.

Pupil teachers were drawn predominately from the hard-working, respectable and religious working classes, but by the final quarter of the nineteenth century some also came from less well-off middle-class families, like Jessie's. Neither she nor her parents were deterred by the working-class setting that discouraged many other GPDSC pupils from entering elementary school teaching and when she began her training in 1886 at Warple Way, a south London elementary school, she was one of the few who had had some secondary education. Her sister, Grace, was still at school when Jessie started receiving a small salary. In 1888 girls earned between 3s. and 10s. per week, and boys between 5s. and 16s.[2] In return, they agreed to undertake a four-year apprenticeship. Jessie's family could offer her support throughout this time and they could see that in the longer term they would all benefit. As a pupil teacher Jessie was now able to contribute to the family income and she could see an independent future for herself, but in the meantime, both socially and academically, she had taken a step down.

When she took up her apprenticeship, Jessie was aged about 17 but many pupil teachers were younger and, at just 13 or 14 years old, you might be little more than a child yourself. You were caught between worlds. You were regarded as the best of the working class and you had been selected to teach others of that class, the children of working men and women. You were expected to act as a beacon to family and friends. Yet your textbooks exhorted you to adopt moral, religious and intellectual habits that were more typical of middle-class adults. You were told you should 'avoid everything that is repulsive, even to the most sensitive, either in manner or conduct'. Vulgarity, rudeness or dirt, and a slovenly appearance were degrading, and 'tobacco and snuff are appropriate only to the ale house or bar-room'.[3] You were required to show firmness, kindness and cheerfulness at all times, and as if that were not demanding enough, self-denial, discipline, diligence and punctuality were also urged.

Not all pupil teachers managed these tensions successfully; some were absent without leave and were subsequently removed from their posts. Some found more subtle ways to resist control of their private lives, their dress and behaviour, and as a consequence found themselves described as sullen, untidy or deceitful. Female pupil teachers were more likely to experience pressure to conform and in particular, their contacts with men were carefully monitored. Jessie, however, was well-regarded at Warple Way and her conduct, both in and out of school was 'all that could be desired'. A testimonial from the school described her as truthful and conscientious, her dress and general demeanour as very modest and unobtrusive.[4] She did not want to jeopardise her future or create financial difficulties for the family; she had social and educational advantages over many of her fellow pupil teachers, and she shone.

During her training at Warple Way School, Jessie also attended a local centre for pupil teachers set up by the London School Board. Here, she mixed with other pupil teachers and followed a curriculum that would soon become very like that on offer in the high schools. Drawing had practical importance since pupil teachers needed to learn how to reproduce figures and diagrams on the blackboard. The other compulsory subjects taught in London centres were Scripture, Arithmetic, Mathematics, English, Geography, History, Science, French and Music. After three years at Croydon High School for Girls, several subjects (which were entirely new for pupil teachers from the elementary schools) were familiar to Jessie. As well as the required subjects she passed government examinations in Physiology, Chemistry and Art. These courses attracted grants from the Science and Art Department and were often merged with the pupil teacher curriculum.

But the classroom was the real test. D. H. Lawrence was himself a pupil teacher and drew on his experiences in his novel *The Rainbow*. When the character Ursula faces her class for the first time, she is overwhelmed: 'It made her feel she could not breathe: she must suffocate, it was so inhuman. They were so many, that they were not children.'[5] Helen Corke also began her teaching career as a pupil teacher. She was a few years younger than Jessie, but followed a similar career path. When she joined the junior department of the local Whitehorse Road school as an apprentice, she found about 80 girls and boys in her class, aged about eight to ten, 'clean and well-brushed, dirty and ragged, who sit with elbows touching, seven or eight in each long desk, holding slates'.[6] Like the fictional Ursula, she, too, fought to keep order:

> can what I am teaching sustain the interest of the class? Eyes must *not* wander, feet must *not* fidget. What will happen if the teacher is away more than a few minutes? Why should these children obey *me*? They don't want to obey – they want to run out of those desks, to shout and dance round. They *must* not – I should be shamed and discredited if they did![7]

Life on the bottom rung of the school hierarchy could be taxing and a pupil teacher might struggle to survive. In the late 1880s Jessie was amongst those who successfully completed the four-year apprenticeship and she passed the final scholarship examination. Two more years of study were required to become a fully qualified teacher and some of her classmates who had good scholarship results went to university now. Others took evening classes. Jessie opted for a full-time course at a residential training college and used the scholarship to subsidise her fees. Corke had fewer choices when she completed her pupil teacher training a few years later. The fees for training college (though less than for university) were beyond her reach. About half of her classmates would go to a training college and she is contemptuous of the way they will spend their time,

> Not in a university, with the opportunity of making social contacts which should widen their experience of life, but in a government subsidised boarding establishment where they will meet only the trainees of their own profession, and where lectures given will relate to their work as elementary school teachers.[8]

Nothing in the records suggests that Jessie felt disdain for her college peers. Perhaps she considered university at this point in her life but, in contrast to Corke, saw value in the college training – and there was her family to think about too. The Corke family had fallen on hard times, and Helen Corke resented their reduced circumstances, but the Cash family were steadily bettering themselves. Orders for quality stationery at Millington's had increased and James' income was growing. Grace had just left Croydon High School for Girls and was hoping to go to Cambridge. Henry was still in full-time education. Kate was already employed as a kindergarten teacher and helping to supplement the family's earnings. If Jessie lived away from home this would also take some pressure off the budget. So, two years after completing her apprenticeship she took the next step towards a teaching career. The Reverend Jupp and her doctor, as well as senior staff at Warple Way School, all warmly supported her application and, in 1891, after paying an entrance fee of £20, she became a resident student at Stockwell Training College in south London.

The college had been founded some thirty years earlier by religious Nonconformists, and when Jessie arrived it was located on the east side of Stockwell Road. Nearly 70 young women entered the college with Jessie in 1891 and many, like her, again took further government courses; Jessie now added Advanced Physiology and Botany to her portfolio. She completed her teaching certificate (with a Division Two pass) and left Stockwell College at Christmastime in 1892. The following year, at the age of 24, she took up her first post at Mitcham Road Girls' School in Croydon and was Assistant Mistress there for the next eight years. 'Ambition, for the assistant mistress', observed the unhappy Helen Corke, 'could only take one form – the desire for a headship.'[9] She herself had no such desire but Jessie was keen to further her career and in March 1901 she became headmistress of another Croydon girls' elementary school, this time in Sydenham Road. She was still living with her parents but the family had now left Thornton Heath for the more desirable district of Coulsdon. At £100 a year, Jessie's earnings must have been a help. In a similar way, Corke's teaching salary later enabled a move for her family.

There was a cluster of schools in Sydenham Road and under the previous head the Girls' School had been in particular difficulties. An official report found 'a want of general guidance and control on the part of the head teacher'[10] but a few months after Jessie took charge improvements were observed. The girls were now 'in good order', and instruction was 'for the most part satisfactory'. Soon after this, Corke joined the Sydenham Road Junior School as a pupil teacher and in 1905 she finally crossed paths with Jessie. When the Junior School closed, a reorganisation brought Corke briefly onto the older woman's staff.[11] By now, the Girls' School was flourishing; Jessie was earning £130 per year and there were 418 girls on register.[12] But as she neared the end of her time as a pupil teacher, Corke was struggling with ill health and stress. Her final scholarship examination was looming and several times her absences were recorded by Jessie in the school logbook. Many female pupil teachers suffered with physical and mental illness as a result of overwork, and years later Corke vividly recollected her ordeal. She managed in the end to pass the examination, and when she moved to another school Jessie endorsed her conduct, character and efficiency as a teacher.

In 1892, the year that Jessie became a qualified teacher, Kate completed her training in kindergarten work. All the Cash sisters had benefitted from the growth of girls' high schools during the 1870s, and now Kate caught another wave of education reform with the emergence of the 'kindergarten movement'. A few years after leaving the Townley House boarding school in Ramsgate, she had enrolled on a teaching course in a local kindergarten in Dingwall Road. The innovative Croydon Kindergarten Preparatory School Company Limited[13] had been set up in 1880 with the assistance of the Croydon School Board (which had earlier supported the establishment by the GPDSC of the Croydon High School for Girls). Under its headmistress, a German émigré called Madame Emilie Michaelis, the first public company for the promotion of the kindergarten system quickly became renowned as a model for others.[14]

The kindergarten movement was founded by Friedrich Froebel in Germany in the first half of the nineteenth century and it arrived in England soon afterwards. Froebel had developed his ideas while working with the Swiss educator, Johann Heinrich Pestalozzi. By the time Madame Michaelis arrived in England she had worked in kindergartens in both Switzerland and Italy and become a keen advocate of Froebel's approach. At the core of the movement was a belief in the need to treat each child with respect, to nurture their distinctive abilities in a happy environment. Play and activity were the foundations for individual growth. Such educational principles raise few modern eyebrows, but at the time Kate began her training in 1889 the Froebelian kindergarten represented a radical break with conventional methods which were both didactic and moralistic. Froebel also emphasised the distinctive contribution women had to make to a young child's education and believed that failure to recognise the value of women's activities was contributing to a breakdown in society.

With Madame Michaelis as its headmistress, the Dingwall Road kindergarten was committed to spreading Froebel's teachings. Shortly after Kate joined it, however, the operation transferred to west London and the Croydon Company folded. In the ten years of its existence over 70 teachers had been trained, but now a London base was sought. Michaelis took the remaining Croydon students with her (there were about sixty, including Kate) and they continued their studies at what would become the Froebel Educational Institute (FEI).[15] Michaelis became the Institute's first principal and in this role she inspired a generation of young, middle-class women. Harold Oakeshott's younger sister, Mary, completed her kindergarten training here, as did Margaret Moore (who would later marry Grace's brother, Henry). Although no supporter of the growing movement for women's rights, Froebel had in effect created a career path for young, middle-class women, and a viable route out of the late-Victorian home. Kate became part of a new profession for the early years' teacher; it was a supportive network of like-minded peers that worked to bring about educational change, despite ridicule from some in the male establishment.

By this time the National Froebel Union (NFU) had been established and was regularly inspecting the quality of kindergarten training, setting examinations for teachers, and actively seeking to discredit those who were working in the field without a proper understanding of Froebelian methods or philosophy. Kate successfully passed her elementary NFU certificate in 1891 and a year later, her higher certificate.[16] She studied a range of subjects in Science, Music (she excelled in singing) and Mathematics. For the higher certificate, knowledge of German was required; other subjects included the principles of Froebel and the theory of education.

And this is where Kate found her niche. A naturally reserved person, she became keenly interested in the history of the kindergarten movement and the philosophy that underpinned it. Eventually, she would become a teacher trainer. When she died many years afterwards, her students affectionately recalled her talks about Froebel and Pestalozzi: 'She kept a favourite picture of Pestalozzi by her desk and after describing his work with children, would turn to it, saying, "And there he is!".'[17]

Kate was committed to the Froebel model for educating young children. It was radically different in approach from methods employed in elementary schools (like the ones her sister Jessie had been teaching in) where a limited curriculum, rote learning and physical punishments were common. Middle-class girls could now follow a 'respectable' route into teaching, one that was not based in a working-class context. Many girls' secondary schools were also offering courses in kindergarten teaching. The GPDSC was keen to promote it since many of their schools had kindergartens attached and they needed trained staff. When, in 1880, Dorinda Neligan became one of the first shareholders of the new Croydon Kindergarten Company, she lent her support to a venture that would enhance career prospects for her senior girls at the high school and she would have known she could guarantee the Company a steady supply of recruits.

Many of the young women who were drawn into kindergarten teaching came with a clear sense of mission. Kate was a lifelong socialist, as well as a staunch Froebel follower, and Michaelis proved a strong, independent figurehead for the movement in England. When she died in 1904 one of Kate's contemporaries wrote about their teacher:

What I best remember of the early student days in Croydon was the simplicity of life, the devotion to work, and the idealism which pervaded everything. I can vividly recall, even after all these years, the feeling of aspiration (exaltation would, perhaps, be a better word) with which we began the day. It was as though one had been set apart for this particular thing, and by its means the world would be regenerated.[18]

Now that both her older sisters were settled into their teaching careers Grace had a decision to make – and in September 1890, at the beginning of her final year at school, an event took place that prompted her to take a different path from either Jessie or Kate. Grace was in the Upper VI form when the first Principal of Newnham College, Cambridge, Miss Anne Jemima Clough, was invited to present the prizes. Hearing Clough describe life at the women's college may well have encouraged and excited Grace, just as it did one of her contemporaries who made the decision to go there after hearing the Principal talk at her own high school.[19] For Grace, however, there was first the question of university fees to resolve, and the following autumn she returned to her school and worked for a year as an assistant teacher. Alongside the experienced Miss Mullins, she taught Reading, Poetry and Arithmetic to junior pupils, some of whom are shown in Illustration 3.1.

Illustration 3.1 This photograph, taken at Croydon High School for Girls (probably in 1891 or 1892) shows Grace (seated on the left end of the middle row) with a class of juniors (including boys[20]) and their teacher, Miss Mullins

At this time, Grace, like both her sisters, may have been actively considering a career in teaching. She would return to the classroom a year or so later but in the meantime, Miss Clough had held out a promise – though formal degrees were not yet available to women, Grace could enter university and pursue her own education further. So, in October 1892, she enrolled at Cambridge. As it happened, she arrived to find that Clough had died from heart disease just a few months earlier and the college was in mourning. A Newnham student wrote of their loss:

> It was what one would have wished for her that she should die in harness among all that she cared for most, and her death drew the members of the college together, as her life had done. All the students of that year felt it an inestimable privilege to be in residence at the time, and they were able to take a last look at the peaceful face of their Principal in her well-known room, among the flowers that her friends had sent.[21]

The death threw a shadow over Grace's early days at the college. University education for women was still in its infancy and its opponents were vociferous, particularly in the early part of the century. Women needed to be taught how to manage children and the home; abstract study was considered an unnecessary indulgence which risked turning them into unmarriageable monsters. It was only after the Second World War that English women finally broke into higher education in significant numbers and so for Grace and her peers, personal encouragement mattered greatly. They were a lucky minority of Victorian women and the support they received from influential pioneers (like Dorinda Neligan, Emilie Michaelis and Jemima Clough) could be critical. They revered and respected the older women's achievements and treasured their association with them, often for the remainder of their lives.

Newnham College had been established in 1871 as a rented house where women attending lectures at Cambridge could live. By the time Grace arrived it had moved to several purpose-built halls on the site it still occupies. It was non-sectarian, unusually for a Cambridge college, and it had no chapel. Grace lived in Clough Hall. She stayed only for one year and, like many of her contemporaries, did not complete a full degree course. The completion of the degree examinations (or the 'tripos') at this time earned women simply a certificate.[22] The knowledge that their achievement would not be formally recognised (in the same way that it was for men) must have been a disincentive for many passionate and able female students. Two years earlier, Philippa Fawcett (daughter of the women's suffrage campaigner, Millicent) had beaten all her male contemporaries by scoring top marks in the mathematics tripos only to have her achievements dismissed as an anomaly.

Students in Grace's time took courses at a level appropriate to their attainment and ability, and individuals whose schooling had not prepared them well for university could begin with familiar subjects and then pursue their interests. Unlike students at Girton, a more radical sister-college, at Newnham you were not required to take the same courses as men and you were not under pressure to prove you were just as capable. In these ways, the college offered less of a challenge than Girton to prevailing beliefs about women's proper role and it met with less opposition.

Grace had missed the chance to know Miss Clough at Newnham but she did meet some other successful women, including the charismatic Mrs Marshall, who taught her Political Economy. As a young Mary Paley, Mrs Marshall had been one of the college's original five students and in 1877 she had married Alfred Marshall. He became an eminent economist but is remembered, too, for having first supported and then vehemently opposed the presence of women at Cambridge. In Grace's time, women were only at Cambridge on sufferance and not all the lecturers were willing to admit them to class. Mary Marshall was one who did:

> Mrs Marshall was as nice as before this morning, but she has given us a fearful paper to write – four long questions and we mayn't read up about them nor will she tell us anything about them till afterwards, we are to answer them from our own Common Sense and by thinking them out for ourselves. Happily she doesn't mind a bit whether our conclusions are right or wrong so long as we *do* think them out to some conclusion.[23]

In contrast to Clough (who according to her niece had 'dressed like a bundle'[24]) Mrs Marshall was both elegant and striking. A student wrote home about her:

> She *is* a Princess Ida. She wears a flowing dark-green cloth robe with dark brown fur round the bottom (not on the very edge) – she has dark brown hair which goes back in a great wave and is very usefully pinned up behind – very deep set large eyes, a straight nose – a face that one likes to watch. Then she is enthusiastic and simple. She speaks fluently and earnestly with her head thrown back a little and her hands generally clasped or resting on the desk ... She looks at political economy from a philanthropic woman's point of view.[25]

Students' working hours were from 9 a.m. to noon, 3 p.m. to 6 p.m. and 8 p.m. to 10 p.m. At these times the Halls were quiet. Grace would have paid an afternoon call each term on Mrs Eleanor Sidgwick (Clough's successor) and in the long lunchtime she would probably have taken a walk. Outside the college, students were obliged always to be decorous and unobtrusive:

> We were asked always to wear gloves in the town (and of course hats!); we must not ride a bicycle in the main streets, nor take a boat out on the river in the daytime unless accompanied by a chaperon who must be either a married woman or one of the College dons.[26]

Restrictions like these were commonplace, of course, in Victorian times, and reflected contemporary attitudes to middle-class women. For many, their college years were actually full of interest and female students often felt they led a much freer life at Cambridge than at home:

> Our lives were excitingly novel. We worked, some of us, ten hours a day, and there were so many College societies and preoccupations there was little time for anything else.[27]

As well as Political Economy, Grace studied Economic History and English History. She did not continue with her school studies in languages and literature but ventured into fields that were new to her. She retained her love of literature, but at Newnham chose to focus on the social and political world of the late nineteenth century and its historical roots. The decision marked the beginning of a formal intellectual engagement with political ideas.

In her English History classes, on Mondays and Fridays, Grace was taught by a Professor Henry Gwatkin, a specialist in ecclesiastical history – and a somewhat less appealing figure in the classroom than Mary Marshall. Gwatkin's hearing had been damaged and his speech impaired by a childhood attack of scarlet fever and communication with him was difficult; he was never comfortable with college life. But he refused to let his disabilities get the better of him and he remained popular at Cambridge. A student who was at Newnham a few years after Grace described him as

> a most original old gentleman whose sight obliged him to hold notes so close to his eyes that his nose almost touched the paper. He also had an engaging habit of walking about among us asking questions and sitting on the edge of tables and swinging his legs.[28]

In Professor Gwatkin's classes Grace was accompanied by a young woman who became a close friend. Renée Courtauld, who was just one year younger than Grace, was also resident in Clough Hall. Unlike Grace she came from a wealthy family. The Courtaulds had made their money in the silk industry, particularly from the manufacture of black mourning crêpe which became popular in Victorian times. Renée was the sister of Samuel Courtauld who would later establish the Courtauld Gallery and Institute of Art in London. She stayed at Cambridge a year longer than Grace but similarly left without taking the tripos. The differences in their backgrounds were considerable, but away from family and familiar social contexts, they drew close. Newnham women were often seen arm in arm and in these sequestered settings lasting friendships were forged. One student later reflected:

> Many of us had lived through lonely schooldays. For the first time, we made friends. The slow exploration of another human being, the discovery of shared perplexities and interests, our delight in our new companions' gifts and, maybe, beauty (for beauty was not wanting in those years) – these were excitements. Many of us made friends who remained faithful to us all our lives. There were other things ... but to me it was the friendship with other girls that I think of, and the springs and autumns of Cambridge with their almost unbearable beauty.[29]

When they left Newnham, Renée and Grace both became activists. Renée was a part-time social worker at the Women's University Settlement, an organisation which sought to improve educational and recreational opportunities for poorer people in London. She was also Honorary Secretary of the Women's Suffrage Society and her sister, Catherine, a member of the Artists' Suffrage League, designed posters and postcards to promote the cause. Renée Courtauld later used family money to support the British Institute in Florence, which had been founded during the First World War to help counter fascist sympathies in Italy. She never married. About twenty years later, in Gisborne, New Zealand, when Grace gave birth to her only daughter, in a poignant reference to an earlier life she named her Renée.[30]

Grace left Newnham in 1893. Although she had no degree and no degree certificate, she had passed a Cambridge examination called the Little-Go (which was a preliminary examination usually taken in the second year). It appears she taught junior pupils at a private school called 'Lammermoor' until Christmas 1895, and when she left there, the children gave her a photograph album as 'a token of their affection and esteem'. The following spring, she returned to Croydon High School for Girls and was again employed, sometimes as an assistant teacher but increasingly she was in sole charge of, for example, drawing classes.[31] At the end of that year, she married Harold Oakeshott and records suggest that she again briefly joined the school staff. Dorinda Neligan was still headmistress and in 1895 she had suffered a 'serious loss' when her secretary, Miss Thomas, left to join an Anglican sisterhood. Grace briefly stood in for Miss Thomas. When she disappeared in 1907 Grace left behind the photograph album she had been given by the Lammermoor children and in it, as well as family photographs, were a couple of small portraits, side by side on the page, of two women who had inspired her. One was Dorinda Neligan and the other, Renée Courtauld.

During the 1890s, Kate, now a qualified kindergarten teacher, had been steadily building her career. She took charge of kindergarten training firstly at the endowed Haberdashers' Aske's Hatcham Girls' School in south-east London and then at two private schools. In 1907, just two months after her sister had vanished, she enrolled for a BA degree at University College, London.

Jessie meanwhile remained headmistress at the Sydenham Road Girls' School until 1906.[32] In November of that year she took leave of absence to join about five hundred other teachers on an educational tour of America, funded by Mr Alfred Mosely, an industrialist who had made his fortune in the South African diamond fields. Mosley was concerned about the threat to Britain from America's greater economic competitiveness. Education across the Atlantic was, he believed, unconstrained by class and religious prejudice and technical training was far superior. It was a system that Britain's practising teachers could learn from.

Jessie was away for six weeks. Just three days before she set sail, an entry in the school log recorded a visit to Sydenham Road by a Mrs Grace Oakeshott. Jessie's younger sister was by now the Inspector of Women's Technical Classes for the London County Council and she had come to observe the method of teaching needlework and the syllabus of instruction at her sister's school. By this time she had become a passionate advocate of technical education for girls and was at the peak of her career. But within the year, she would be gone.

4
'Another Word for Suicide'

In December 1896, a few days after Christmas, Grace marries Harold. Most women of Grace's generation expect to marry although, in socialist circles particularly, the institution is under increasing scrutiny.

When Harold and Grace married in the Croydon registry office on a mild, dry but dull Wednesday in late December 1896, they and their families were setting themselves apart from dominant religious custom. The Marriage Act of 1836 allowed Nonconformists to marry in their own places of worship (such as chapels and synagogues) and it also enabled non-religious civil marriages to be held in registry offices. This was not a common choice in the 1890s but such weddings were relatively inexpensive, private and informal and they avoided all ecclesiastical associations.[1]

If the choice of venue was unusual, the decision to marry in itself was not. Most women of Grace's generation assumed they would become wives and mothers and those who wanted to marry far outnumbered those who did not. Marriage was also 'respectable'. And although most middle-class families believed a college education (like the one Grace had had) was a distraction that would seriously reduce a young woman's chances of finding a suitable husband, it did not deter Harold. The Cash and the Oakeshott families shared a background in radical Nonconformism and both were lower middle class. Harold and Grace were social equals.

Despite the popularity of the institution in the nineteenth century, not all women wanted to marry or found fulfilment when they did. Not all *could* marry either since, as the 1851 Census had revealed, in England women considerably outnumbered men. Neither Jessie nor Kate married and both had successful careers in education. In the 1890s, single women had an increasing number of employment opportunities (in teaching and office work, for example) and in English law Jessie and Kate had more status than their married counterparts. As a single woman you had the same rights to own property as a man, as well as some parish voting rights, and you could act as an agent for another person, for example, as a trustee or executrix of a will. Spinsterhood, despite carrying a stigma, was the logical path for many women who sought independent careers and it could also be a form of protest against restrictive marriage laws. The Married Women's Property Acts (1870 and 1882) had finally given a wife the right to own, buy and sell her property separately, and the Matrimonial Causes Act (1884) meant that if you left your husband, you could no longer be imprisoned. Even so, married women (especially those from the middle classes) remained economically and socially dependent.

What led Grace to put herself in this position? Most Victorian women did not keep detailed, personal records of their private decision-making, particularly in relation to marriage and sex, and without first-hand accounts, their emotional lives are difficult to uncover.[2] One who did record her views and experiences was Beatrice Potter (later Beatrice Webb, the social reformer and socialist) who kept a diary, described by her biographer as a kind of 'other self' to whom she talked compulsively.[3] Marriage, she said, was at best a 'speculation in personal happiness' or at worst, 'another word for suicide'.[4] By this time she had experienced considerable pain and humiliation at the hands of Joseph Chamberlain, an influential politician and business man. His failure to propose marriage to her after her declaration of love had left her mortified and indignant, at times even suicidal. A couple of years later Potter did in fact marry, and by the late 1890s she was, like Grace, a member of the Women's Industrial Council (WIC), a voluntary organisation that campaigned on behalf of working women.[5] In seeking to persuade Beatrice to marry him, Sydney Webb had argued that as two socialists they would achieve more together than apart and Grace may have held similar hopes in relation to her own marriage.[6] Harold was a committed member of several socialist organisations, including the Fabian Society, the Croydon Socialist Society and the Fellowship of the New Life.[7] He was witty and congenial, idealistic and impassioned and like Grace, he was strongly drawn to the socialist cause.

Perhaps Harold and Grace had considered living together without marrying. Like spinsterhood, cohabitation was also sometimes a protest against the constraints of Victorian marriage. In the 1890s a few brave spirits adopted this unconventional way of life but their stories, as Grace and Harold would have heard them, were salutary. Eleanor Marx (the youngest daughter of Karl Marx) had founded the Socialist League with William Morris (a friend and associate of the Oakeshotts). At about the same time, she entered into a relationship with fellow socialist Edward Aveling. They cohabited for fourteen years and openly shunned marriage, describing it as 'worse than prostitution'.[8] It was the most prominent socialist free union of the late nineteenth century and it ended in tragedy.

When he began living with Marx, Aveling was already married to someone who refused to divorce him. Then, when his wife died in 1892, he continued living with Marx but secretly married someone else in 1897. It is not clear when Marx found out about his re-marriage. She had felt bound by the relationship with Aveling precisely because it was based on love alone and so she was especially reluctant to admit failure. In this way, cohabitation was sometimes as constraining as marriage since a breakdown could not be openly admitted without acknowledging the failure of the socialist ideal. Aveling's final betrayal left Marx unable to cope and in 1898 (two years after Grace's marriage to Harold), she committed suicide, leaving Aveling all her money. He was widely reviled and held responsible by many socialists for Marx's death.

Another socialist at this time who refused to observe convention also paid a heavy price. Edith Lanchester worked for a time as a secretary to Eleanor Marx. She joined the Social Democratic Federation and there in the early 1890s she met James Sullivan. They fell in love, but Lanchester refused to marry. She believed marriage turned women into chattels and she made public her intention to follow Marx's example, and to live with but not marry Sullivan. Her family were appalled. In 1895 her father and brothers found a doctor to certify her and she was forcibly removed to a lunatic asylum in Roehampton. The doctor explained that just as he would have certified her if she had threatened suicide, so he was justified in trying to prevent her from committing 'social suicide'. The case attracted enormous publicity and Lanchester was freed after four days when it was decided that she had been detained without sufficient cause. She never married Sullivan – and she never saw her father again.

Were Harold and Grace influenced by what they heard in the 1890s about their fellow socialists, Marx and Aveling? Or about Lanchester? When they considered marriage themselves, did they talk about what sort of union they wanted? For many women, despite the legal restraints, marriage meant freedom from dependence on parents and it offered the chance of making separate friends. Some women in practice gained independence through marrying and were not at all dominated by their husbands. When Grace and Harold first met he was employed as a tea-taster with a London tea specialist. During the 1880s and 1890s tea consumption was rocketing and the import of tea from India and Ceylon (Sri Lanka) was making people rich. It was a prosperous industry and, to a Victorian woman and her family, Harold would have looked like a steady breadwinner. He was good husband material, and with their common faith in the socialist cause, it is easy to understand how the couple might have been drawn to each other.

An Australian-born artist, Stella Bowen, wrote about meeting young socialists in London in the early years of the First World War (not long after Grace had left England). In her autobiography Bowen describes the comrades she met as 'cheerful' and 'selfless'. She observed that they relaxed easily into 'a happy intimacy'. Their dedication to the cause seemed to her to 'make personal relationships unimportant, and when personal relationships are unimportant they are not cultivated with skill'.[9]

We have no record of Harold and Grace's courting and no diaries of their thoughts and feelings about each other. It might have been some time before Grace was aware that Harold had a taste for stronger beverages than tea – and for now the proposed marriage had all the appearances of respectability. Even contemporary socialists and feminists believed that, in the context of Victorian society, marriage could be a sensible decision. And it need not be constraining.

After her marriage, Grace showed little propensity for the traditional wifely role. Instead, she went back to work. In the 1890s, in 'developed' European countries, only 12 per cent of all married women had a recognised occupation outside the home.[10] In the English middle classes it was frowned upon. It was not 'respectable' behaviour. Married women who took paid jobs supposedly deprived both single women and men of employment opportunities and they threw into question their husband's status, even his masculinity. So when Grace returned to her old school to work as a secretary, albeit briefly, she challenged the prevailing orthodoxy. Perhaps she had been asked as a personal favour to help the headmistress, Dorinda Neligan, whom she had come to admire. Or perhaps the post was temporary and children were expected to appear soon and keep her fully occupied. In any event, she had time now to plan her future, and to make the personal contacts she would need to pursue her political goals.

On some occasions, both before and after her marriage, Grace accompanied Harold to socialist meetings. Two women, Clementina Black (a friend of Eleanor Marx), and Amie Hicks, both of whom would soon become her colleagues on the WIC, addressed Fellowship meetings in the 1890s. They spoke about working conditions for women and girls in factories, about trade unions for women, and about the 'ethics of shopping' (the need for consumers to pay a fair price for their purchases). Mona Caird, an active suffragist and feminist writer whose views were creating controversy in the late nineteenth century, was also an early contributor to Fellowship meetings, where what was becoming known as 'The Woman Question' was frequently discussed.[11]

Campaigns for women's rights to education, property and the vote were escalating in the 1890s. Caird believed that women ought not to be tempted to marry or to stay married simply for their bread and butter. Such marriages were an insult to human dignity. The ideal marriage would be free and the first condition was that women should be economically independent:

> we look forward steadily, hoping and working for the day when men and women shall be comrades and fellow-workers as well as lovers and husbands and wives, when the rich and many-sided happiness which they have the power to bestow one on another shall no longer be enjoyed in tantalizing snatches, but shall gladden and give life to all humanity.[12]

By now, the term 'New Woman' had entered the popular vocabulary and become journalistic shorthand for women who, like Grace, sought greater autonomy and believed in the need for social and political reform. Commonly depicted on her bicycle, or smoking a cigarette, the New Woman was an icon of the 1890s and a challenge to Victorian images of femininity. She was mercilessly mocked in the media – but in the novels of writers like Sarah Grand she became a vehicle for promoting social and sexual change.[13]

This emerging genre of New Woman fiction often explored married women's experiences, and many of the stories were a grim indictment of the institution. The Victorian novel conventionally ended with wedding bells and progeny, as in George Eliot's *Middlemarch* for example, but now the tone darkened. In a short story, *The Yellow Wallpaper*, written by Charlotte Perkins Gilman and published in 1890, the narrator documented her physical and mental collapse following her marriage and the birth of her first child. Her husband John, a physician, had forbidden her to work. She must rest and focus exclusively on domestic life. Just as Gilman herself came to challenge the 'rest cure' prescribed for her following her own breakdown, the narrator in her story believes that congenial work, with excitement and change, would do her good:

> I sometimes fancy that in my condition if I had less opposition and more society and stimulus – but John says the very worst thing I can do is think about my condition, and I confess it always make me feel bad.[14]

The year after Gilman published *The Yellow Wallpaper* she wrote an essay called 'The Extinct Angel', in which she likened 'the angel in the house' to a dead dodo. As Grace grew into adulthood, in radical middle-class circles, the Victorian ideal of a wife, selflessly devoted to children, passive and refined, sympathetic and charming, was under attack.

Childbearing was central to this ideal of womanhood. It brought fulfilment to a marriage and it kept women at home. Any rejection of the maternal role was a challenge to the Victorian order. But Grace had energy to spare and a commitment to social action. Without children she could spread her wings. In 1897 she left the secretarial post at her old school and soon afterwards, joined the WIC. For the next ten years she worked as an unpaid investigator, visiting factories, talking to employers and women workers and reporting on their conditions. For a married middle-class woman, voluntary work like this was more acceptable socially than a paid job. And Grace had Harold's backing. The WIC collected donations from its supporters including, in 1898, both Harold and Grace. At about the same time the couple moved to Coulsdon in Surrey, to a large house Harold had built for them, in Fanfare Road.[15] With its nine rooms and thirteen doors, 'Downside Cottage', shown in Illustration 4.1, was comfortable and had ample space for children – but still none appeared.

Illustration 4.1 Downside Cottage in Fanfare Road (later Downs Road) in Coulsdon, Surrey, where Grace and Harold lived after their marriage. Grace's parents, along with Jessie, Kate and Henry also moved from Thornton Heath and lived a few doors away in the same street

In fact, the absence of offspring was likely to have been a deliberate and conscious choice. Sometime in the late 1890s Harold put forward his views on marriage at a meeting of the Fellowship of the New Life.[16] A conventional marriage results all too often, he said, in a narrowing of interests. In a life lived to 'a tolerably easy standard', each looks after the comforts of the other, and tries to spare them everyday drudgery. However, there is a wider social cost. A husband will typically take to gardening or golf and neglect his political activities and colleagues. A wife will focus on acts of local charity and neighbourly kindness. 'But', said Harold, 'any suggestion of a radical cure for the poverty she is seeking to alleviate is too fraught with danger to her own security to be entertained.' Improving the situation for the worse off members of society, in other words, comes at a cost for the better off and it is a cost not all are willing to pay.

Grace was present at the meeting and now heard Harold voicing his support for her political aspirations. She would not settle for local acts of kindness, or be distracted by motherhood. Instead, she would seek to bring real change and concrete benefits to the lives of working-class families, and clearly she had her husband's approval. Children, Harold went on to say, were often produced through a desire for immortality, as 'a cry of despair' when the married couple failed to find satisfaction in their own lives. Of course there was social value in bringing up children who were well equipped for 'the struggle' but greater social benefits would result if the individual's resources and energy were used to improve the environment for all. In their efforts to obtain good schooling for their children, for example, parents were left with no time to agitate for better education for all. They felt obliged to live in high rent neighbourhoods and so could not afford to make a meaningful contribution towards the development of the poorer neighbourhoods from which they had escaped. Generally, said Harold, it was those without children who worked hardest for the common good.

In his talk Harold put emphasis on the value of the individual life. Love, truth and freedom were the keys to an improved society – not 'blind obedience' to institutional practices and constraints. Even a few months after the marriage, Harold's report on a Fellowship Gild he had set up for local working-class youngsters suggested a fondness for young people.[17] Indeed, he might have wanted children but, at this point in his life, been willing to settle for friendship and respect, based on a common socialist purpose. And although Grace's work with the WIC was unpaid, it soon absorbed most of her time and brought her substantial secondary gains. She came eventually to have status and influence for the first time in her life, the companionship of like-minded women, and access to policymakers. This was a marriage that would both support and free her.

The options for couples (like Harold and Grace) who wished to remain childless were limited. Nevertheless, in the late nineteenth and early twentieth centuries, the number of childless marriages was increasing and overall births were falling. In less than sixty years the total fertility rate was more than halved.[18] More European women were marrying at a younger age, but still the birth rate dropped – and the inescapable conclusion was that deliberate birth control had spread.[19]

Birth control at this time overwhelmingly meant abstention from sex or *coitus interruptus*. By the early twentieth century both these practices were frowned upon as they were thought to cause anxiety and other nervous conditions, especially amongst women. Despite such advice, family size continued to decline even amongst the fittest and those most able to reproduce. Doctors and clergymen often blamed middle-class women for the falling birth rate and accused them of being 'selfish'. Eugenists were dismayed because they believed more births were needed amongst those with social standing (who were assumed to have genetic worth). Only in this way, they argued, could the fitness of the 'race' be assured.[20] Nothing in Harold's writings suggests that he subscribed to eugenic beliefs. He mentions them in passing and then sets them aside to stress instead the importance of each individual's life. Some socialists did believe in the need for 'scientific planning' to create a better society and several prominent Fabians came to support the controversial doctrine, amongst them Harold's associates, George Bernard Shaw, H. G. Wells and Sidney Webb.

Throughout the nineteenth century, as birth control increased, a new ideal of the 'companionate marriage' emerged amongst the progressive middle classes. The term was used to describe a partnership based on 'mutual restraint, forbearance and respect'. It was a comradely union, a 'higher form of conjugal companionship'.[21] In a companionate marriage the 'reasonable' husband, despite his legal rights, would not insist on sexual union. Harold and Grace knew of other childless marriages, like those of Sydney and Beatrice Webb, and George Bernard Shaw (whose marriage to Charlotte Payne-Townshend remained unconsummated). Her family would say, years later, that Grace's marriage to Harold was also unconsummated. Though they might, like the Webbs (who indulged in what Beatrice Webb quaintly termed 'spooning'[22]) have had a limited physical relationship of some kind, what is certain is that no children resulted, even though both were capable of procreation, as later events would show.

Sexual abstinence entailed a personal cost, of course, even for those with shared political beliefs in its necessity, but whatever bargains were struck in the bedroom, Grace and Harold remained together for eleven years in what must have been a stable and companionable enough arrangement. And the couple would have understood how difficult it was to escape a marriage contract in the early 1900s. In 1857 the Matrimonial Causes Act for the first time made divorce available to ordinary people (and not just rich men who could afford to pass a private Act of Parliament). It was, however, prohibitively expensive and it could not be consensual; it had to be contested by one of the parties. To obtain a divorce a husband needed to prove only his wife's adultery, but as a wife Grace would have needed to prove both adultery *and* a further cause, such as desertion, cruelty, incest, sodomy or bestiality – and there is no indication that any of these things formed part of the marital relationship.[23]

As well as the high cost of obtaining a divorce, Victorian couples had to consider the likely damage to their public standing. The divorce court, set up by the 1857 Act, had come to be widely feared amongst the respectable middle and lower middle classes. Throughout the last half of the nineteenth century, those regarded as less respectable (bankrupts, publicans, drunkards and seducers) all used the divorce court without much concern for the shame or scandal that might follow, but others went to considerable lengths to avoid it. By the 1890s it was clear that litigation could ruin reputations and bank balances alike.

Contemporary socialists were not immune to these concerns. In his Utopian novel *News from Nowhere* (first published in 1890) William Morris imagined a future in which there was no contractual marriage and no 'such lunatic affairs as divorce-courts':

> We do not deceive ourselves, indeed, or believe that we can get rid of all the trouble that besets the dealings between the sexes. We know that we must face the unhappiness that comes of man and woman confusing the relations

between natural passion, and sentiment, and the friendship which, when things go well, softens the awakening from passing illusions: but we are not so mad as to pile up degradation on that unhappiness by engaging in sordid squabbles about livelihood and position, and the power of tyrannising over the children who have been the results of love or lust.[24]

Some comrades were devising their own solutions to these dilemmas. As well as the Fabian Society, Harold had joined the Independent Labour Party (ILP) and both these organisations were accused of sexual hypocrisy. Although Harold was not one of them, there were some individuals whose practice did not sit comfortably with their principles. One infamous episode concerned the writer H. G. Wells, who joined the Fabians in 1903. His polemical fantasy *In the Days of the Comet* appeared in print in 1906 (the year before Grace disappeared) and caused consternation amongst his fellow Fabians, particularly the established membership or the 'Old Gang'. Joseph, Harold's older brother, was by now a member of the Fabian Executive, and undoubtedly part of this group. Both he and Harold would have witnessed the ensuing controversy at first hand. Like the earlier *News from Nowhere*, Wells' novel also put forward a radical vision of a utopian future. There was to be no marriage and no sexual taboos. In the luminous green and gaseous tail of a comet, the earth is cleansed, and relationships amongst its inhabitants are altered morally and fundamentally. Free love reigns in a world without sexual jealousy, and everyone is reborn.

Wells was already out of favour amongst some Fabians for rebelling against the dominance within the organisation of Sydney and Beatrice Webb.[25] His readiness to talk openly about his liberal beliefs in relation to sex and marriage (and more particularly his willingness to live by them) horrified his colleagues. He had left his first wife to live with his second, and he had numerous affairs during this subsequent marriage, including one with twenty-year-old Amber Pember Reeves (daughter of fellow Fabians William and Maude Pember Reeves) which led to a public scandal.

Wells found the hypocrisy of his Fabian colleagues maddening. They talked easily about equality and freedom but seemed unable to tolerate a lifestyle based on these values (as he believed his was). For their part, his critics believed such behaviour would bring them into disrepute with the working-class labour movement and alienate the very people they sought to influence. The Fabian Society was not committed to free love. Respectability was important to the advancement of the Society's political agenda and they could not afford to be associated with sexual promiscuity or with behaviour that undermined traditional marriage. The ILP adopted a similar stance when Tom Mann (ILP secretary in the 1890s) left his wife to live with Elsie Harker, an opera singer and Labour activist. Mann and Harker went to live in Australia to escape the scandal. They were never able to marry since Mann's first wife would not divorce him, but like most working-class cohabitees, they pretended to be married even when they were back in England. Mann never re-joined the ILP.

Socialist groups that were seeking to establish themselves and attract wide support were not willing to tolerate marital nonconformity. It was not a vote winner. Less established sections of the labour movement, like the Fellowship of the New Life, were more accepting of unconventional unions but socialists who (like Harold) belonged to several different organisations might find themselves uncomfortably positioned in relation to these debates. And as a Victorian woman Grace was even more vulnerable. A double-standard prevailed. The opposite of the Victorian 'angel in the house' was the 'fallen woman', and as a female icon it was just as potent.

5

'Fellowship is Heaven'

The story returns now to the 1880s and the years before Harold's marriage to Grace. 'The Fellowship of the New Life' is flourishing in Thornton Heath. Then in the late 1890s, Henry Cash, a keen yachtsman, sails off the coast of England. On several occasions, he is joined on board by Walter Reeve and later, by Grace and Harold.

If there was to be no more contractual marriage, as William Morris had advocated in his nineteenth-century Utopian novel *News from Nowhere*, then what would the new social order look like? What was the perfect future the socialists were seeking? William Guest, the novel's narrator, is fretful about it: 'If I could but see a day of it', he said to himself, 'if I could but see it!'[1] After a fitful sleep, he wakes to find himself in the early twenty-first century, when the River Thames teems with salmon, the local waterman does not expect to be paid for his services, and handsome buildings create an 'exhilarating sense of space and freedom'.[2] People no longer suffer the burdens of industrialisation and there is no formal schooling for children. Instead of marrying each other, men and women are free to move about and live in groups of various sizes, as they please. This is a new day 'of fellowship, and rest, and happiness'.[3]

Morris died in 1896. Shortly afterwards an appreciation of his life appeared in a journal called *Seed-time*, published by The Fellowship of the New Life. The writer signed himself with the initials 'JFO' – and he was Joseph Francis Oakeshott, Harold's older brother. The organisation had, he said, lost a friend. Morris had inspired them with his belief in the supreme value of fellowship and his ideal of a noble, simple life:

> For where amongst modern men and women could one find a better example of an all-round man: full of varied but co-ordinated sympathies; disgusted with the low ideals and the lower lives that he saw around him

and always yearning for a better time to come, yet determined to make life now as full and beautiful as it could be? No man seemed to be so full of the joy of living, or so able to inspire one, whilst with him, with a similar joy.[4]

Joseph's tribute captured the spirit of the Fellowship. They longed for a better life, and not just for themselves. Members struggled to think and act in ways that would help others. How could the existence of ordinary people be improved? Their inspiration came from the teachings of a Scottish philosopher, Thomas Davidson. In October 1883 a circle of Davidson followers had met in Regent's Park, London, to talk about his work, particularly a recent paper of his entitled 'The New Life'. Intense discussion about the best way to live and to improve welfare and happiness for everyone soon resulted in a split. There were those who believed spiritual considerations must be paramount. The way forward was to focus on education, simple living, manual labour and religious communion. These individuals now formed The Fellowship of the New Life, whilst those who wished to devote their energies to the pursuit of material and economic reforms, to be part of an intellectual pressure group rather than a communitarian experiment, soon constituted themselves as the Fabian Society.

After the split with the Fabians, the Fellowship became firmly based in Thornton Heath, and Joseph moved there from north London to join like-minded enthusiasts.[5] Several individuals (including the Oakeshott brothers) eventually became members of both the Fellowship and the Fabian Society and Joseph (a civil servant with expertise in statistics) was influential as a writer of several Fabian tracts.[6] He and Maurice Adams kept the Fellowship alive for the next thirteen years, with their editing, lecturing, writing and organising.

The Reverend William Jupp was another early Thornton Heath activist.[7] Like others in the Fellowship, he believed that the frenetic pace of nineteenth-century urbanisation had alienated people from the natural world. There was perfection in nature, 'But in human life, with its civilisation and culture', he argued, 'how rarely if ever is it so. Here it is the imperfect and the unfulfilled that confronts and so often baffles us.'[8] Closeness to nature might free the spirit from the encumbrances of town and city and foster a truer appreciation of our place in the universe. In middle-class suburbia, in Victorian Thornton Heath, for example, a rural life could at least be imitated.

Others in the Fellowship were more concerned with the social world and keenly deplored the living conditions that many poorer people endured. Adams was editor of *Seed-time*[9] and when it first began publication in 1889 he called for a freedom that was not based on the exploitation of others. A simple as opposed to a 'fashionable' life with all its 'unloveliness and folly', a dissolution of class barriers, an enjoyment of beauty, the community of others – members of the Fellowship would work to exemplify these values in their own lives because, claimed Adams, 'fellowship is heaven, and lack of fellowship is hell; fellowship is life, lack of fellowship is death'.[10]

Fellowship values contained an appealing blend of ideas, a sort of fluffy theism mixed with ethical socialism.[11] The organisation would later be described by Jupp as 'a kind of church or communistic brotherhood'.[12] Although its proclamations had clear religious overtones, its members wanted little to do with conventional theological doctrine. For them, the key to becoming better citizens of a better world was community – and, wrote Joseph, a kind of social, practical Christianity in which the importance of this world was greater than that of the world to come.[13]

At Oak Villa in Beulah Road, some of the Fellowship's members lived communally. A kind of collective living had been discussed from the very beginning and many members thought their ideals could only be put into practice if they created communities or colonies. In Nonconformist circles at the time, families and couples often relocated to be near like-minded progressives and those with whom they shared a radical outlook. The clustering of Fellowship members and their families in Thornton Heath from the mid-1880s was a loose and informal attempt to go beyond this; not to retreat from the world in any way, but to realise a vision and give substance to a new imagining of human life, based on love, truth and freedom.

The Fellowship established their own ethical church, ran a small printing press, a kindergarten, and a programme of talks and discussions. As well as regular meetings on Sunday mornings they hosted social events, an 'historical reading class' and a debating society. On one occasion Grace's father, James Cash, opened a debate on 'Is the family declining?' In 1889 he could be found reading a paper on the 'New Industrial Organisation',[14] and in that year the programme also included lectures on 'the spiritual principle', ethics and religion, and freedom of the will.

It was heady stuff for a Sunday morning. Croydon was full of small political and religious groupings, some more eccentric than others, and when Harold arrived from Sunderland in the early 1890s, he was immediately caught up. He joined forces with his brother and his sister, Violet (who was also by now active in the Fellowship). Harold attended Fellowship meetings on Sunday mornings; probably, at least in the beginning, on Sunday afternoons he was at the Croydon Brotherhood Church (a diverse ethical organisation interested in social and political reform),[15] and then on Sunday evenings he was with the Croydon Socialist Society, where he was a member of its first Committee. It was a lively time, full of intense excitement and optimism. Socialism was a way of life for its disciples, not just a topic of conversation in pubs and clubs. Men and women alike joined the movement and they did not come from just from one social class or age group or region. Years later Nellie Shaw, who was Harold's colleague at the Croydon Socialist Society and another early enthusiast for ethical socialism, remembered the crowd at the Brotherhood Church:

> Every kind of 'crank' came and aired his views on the open platform, which was provided every Sunday afternoon. Atheists, Spiritualists, Individualists, Communists, Anarchists, ordinary politicians, Vegetarians, Anti-Vivisectionists and Anti-vaccinationists – in fact, every kind of 'anti' had a welcome and a hearing, and had to stand a lively criticism in the discussion which followed.[16]

In contrast to the Brotherhood Church, the Fellowship's membership was solidly and respectably middle class, but by the time Harold joined the organisation was seeking to broaden its influence and to bring more people under its sway. Often the members found themselves debating how they could achieve this. Then, at a picnic in the nearby woods of Croham Hurst one day, somebody raised the idea of extended social action, and a kind of outreach project began to take shape. The Fellowship already ran its own kindergarten but the older children needed classes too, once they had outgrown kindergarten games. So, in 1893, a separate 'Gild' was formed and Harold took the lead.

Initially, members proposed a colony, with a meeting hall and cottages to let. Here, they might choose to live in fellowship with one another. Building schemes soon gave way to suggestions of using a mansion or other vacant premises, as Harold explained:

> At last, after most of us had acquired a life's experience of house-hunting, we pitched upon a far from palatial little shop in South End, which offered, in addition to the shop, two decent rooms, a kitchen, and two villainously low attics. Four volunteers for immolation on the altar of Fellowship engaged rooms and everything, we thought, was settled except a small fraction of the rent and the landlady's agreement to the alterations we required. When – bitterest blow of all – not only would she not agree to our terms, but refused to accept us on any terms.[17]

A little hall was found in the end, and soon a handful of children had gathered. The Gild was to be run on democratic principles and as far as possible the children themselves would decide the activities. 'As we imagined, they required suggestions', says Harold, 'and so we were enabled to get them of their own free will to select the only classes we were in a position to supply.'[18]

Soon, there were classes in drawing and painting, sewing, cooking, reading, musical drill, singing and dancing for the girls and gymnastics, games and chip-carving for the boys. Later on, efforts were made to capture the parents. When workers suggested to the older boys that they might like to bring their parents to one of their meetings, writes Harold, 'the universal yell of derision with which the proposal was met showed a link or two missing in the paternal and filial relationship'.[19] But the Fellowship members persisted:

> It was on one of these evenings, as some of us were waiting for the adults to turn up, and – shall I own it? – hoping they wouldn't, that the door was thrown open and we saw a pleasant-faced youth standing somewhat diffidently in the opening. A quick movement on either side showed him to have been pushed forward as spokesman for others whose modesty kept them out of sight. He doffed his cap, took a step or two forward, bowed, and said 'Good evening', adding persuasively 'May I come in?' With the desired permission he bounded out and returned with two others who had been awaiting the verdict outside. The three showed themselves obliging to an embarrassing extent – they would do anything we liked, but they wouldn't presume to suggest.[20]

Harold was well disposed towards the local youngsters. Later, he would become an affectionate grandfather, and he is still remembered for his entertaining ditties and quirky improvisations. In the late 1890s the Gild took up at least one of his evenings a week and it continued to be managed by its young members as far as possible.

Grace had returned from Cambridge in the summer of 1893 and may well have lent a hand, for example with organising the girls' classes. The children in the Gild subscribed to no creed, said Harold, and the aim was simply to provide them with enjoyment and opportunities to use their leisure constructively. All the same, he was serious about the Fellowship's underlying purpose and committed to it. He wrote and talked passionately about things he held dear. The free enjoyment of nature was more important 'than a big balance in the bank'.[21] A consideration for the feelings of others and 'an individuality which radiates without oppressing' – these were the qualities that mattered.[22] He had been first attracted, he said, by the Fellowship's insistence on a simple, full and rational life, 'as opposed to the life cramped by unreasoned conventions and superfluities'.[23] And both he and Grace believed strongly in social action. A full individual life could only be achieved by those who took 'some active part in the work of the world'.[24]

The Fellowship would eventually come to be regarded as the wellspring for much of the ethical socialism that was spreading throughout England at the turn of the century and both Grace and Harold were swept up by its optimism and zeal. When the organisation folded in 1898 Fellowship members believed their work was done. Although they could not claim to have converted the world to their principles and ideas, they were no longer a lone voice in the wilderness: 'What we whispered in the ear in closets is now being proclaimed from the housetops,' wrote Maurice Adams.[25] Now Harold hoped their teachings would be spread through 'the lives of those who have received the seed of the Fellowship ideal',[26] people who, like him, believed that self-realisation ranked above uniformity, sectarian dogma and social convention. He had been married to Grace for just over a year.

How do you live a life like the one Harold describes? How do you live outside 'social convention'? Did he think of marriage as mere social convention? And what does 'self-realisation' mean in the context of wanting improved conditions for everyone? Others in the Fellowship were also debating these questions in the 1890s. The organisation remained centred on Croydon where its most active followers were based, but some London members had earlier begun another and more formal co-operative experiment. In 1891 the Fellowship had purchased a house at 29 Doughty Street in London's Bloomsbury district and Edith Lees (who later married Havelock Ellis, the writer and sexologist) took charge of the new venture, together with Ramsay MacDonald (who in 1924 would become the country's first Labour Prime Minister).

Lees seemed to be the embodiment of the confident New Woman.[27] For her, the only way to bring about change was through action. Women would be chiefly responsible for the advent of the new life, she said: 'the only way to bring the ideal into the real – *is to do it.*'[28] But 'doing it' proved difficult and the Doughty Street experiment was short-lived. Lees resigned a year after it had begun, claiming in exasperation that, far from being heaven on earth, 'Fellowship is Hell'. Later, under her married name, she published a novel called *Attainment* which was based on her experiences in the Bloomsbury commune.[29]

The story follows the fortunes of a young Rachel Merton, daughter of an enlightened village doctor. After working in London for a philanthropic organisation, Rachel falls under the spell of a charismatic poet and is persuaded to help form a new communistic society. Its aim is to live an ideal life – true, real and completely uncompromising. Men and women will join together and live under the same roof, in a comradely fashion. Celibacy is discussed – would such a requirement help to keep undesirables away or would it tax human nature too far? And what should they call themselves? Are they modern crusaders, glow-worms, lanterns, pioneers or rebels? Lees makes a playful allusion to the Fellowship itself when her protagonist Rachel suggests they might call themselves 'the sowers'.

Eventually, however, they settle on 'The Brotherhood of the Perfect Life'. But does it smack of priggishness, they wonder? Sternly, the group are reminded that they will be apostles of a new life – and that no apostle or follower of a spiritual leader minds ridicule. But within three years the foibles and eccentricities of various fictional members, the outside world's incomprehension and disapprobation put paid to the venture and Rachel leaves. At the end of the novel she declares herself a convert instead to a more fulsome love of humanity; brotherhood was frankly an experiment she says to her father. Real brotherhood is for all times and all people and cannot be limited by rule or communities. Rachel must be free to be herself.

Lees may have found her experiences at Doughty Street less than uplifting, but elsewhere other utopian projects continued to spring up. The aim was to come together, to be better socialists and to enlist others to the cause. It was a frenetic, intoxicating time. The movement was itself the means by which belief and commitment could be strengthened and it was more like joining a cult or a sect than a church or a political party. Two utopian projects that had their roots in the Croydon Brotherhood Church in the last decade of the nineteenth century were the Purleigh Colony in Essex and Whiteway in the Cotswolds, in Gloucestershire. The colonists at Whiteway followed Tolstoy's teachings and thought of themselves as anarchists. Many local socialists were deterred by this emphasis and the Oakeshotts would have been amongst them. Joseph advocated law reform and political change – but not anarchy. Harold was by this time aligning himself with the organised labour movement and he, too, would have been unsympathetic to Tolstoy's suggestion that one should not occupy any Government job or employ the law in any form.[30]

Nellie Shaw describes how the Whiteway colonists at the end of the nineteenth century used to make their own entertainment:

> Possessing an excellent piano and an almost more excellent pianist, we have some most enjoyable times. Some of us sing a little – solos, duets and part songs – and we hope to do more in that way as the long evenings draw near.[31]

They also read passages of prose or poetry aloud to each other. And while the colonists in Gloucestershire were creating their own amusements, in townships like Croydon entertainment was rapidly becoming more organised. By now you could, for example, attend orchestral concerts, like those at nearby Crystal Palace that Helen Corke went to on Saturday afternoons, for just a shilling. In her autobiography Corke recalls her joy on first hearing the music of Handel, Haydn, Bach and Beethoven.[32] She also delighted in the Surrey countryside, where she took leisurely rambles with her companion, D. H. Lawrence. On the North Downs, the pathways quickly took you away from the crowd and you could enjoy nature.

Grace, too, had discovered the remote beauty of the North Downs. Her favourite retreat was Ranmore, a steep-sided common near Dorking. It was some distance from Downside Cottage in Coulsdon where she and Harold lived after their marriage but you could go part of the way there by train. Downside Cottage was itself a hide-and-seek maze of rooms and doors, offering plenty of scope for independent living. It was easy to get outside into the garden. On one side there was a bank with narrow paths winding from top to bottom. In summer, nightingales sang from the trees and the local harebells grew high.[33] Still today, Farthing Downs are just a few steps away, and as you cross the road and walk up the grassy slope, the house quickly escapes from view.

Grace's brother also took pleasure in the natural landscape. Henry was a keen yachtsman and in the summer of 1895, Walter Reeve (now aged about 19) joined him for a weekend on board a hired cutter called *Scout*. It was the first time he had crewed for Henry but he took easily to the task and the two young men enjoyed each other's company. On the Saturday, they sailed from Burnham-on-Crouch in Essex, out into the wide estuary, and eventually they dropped anchor for the night off the east end of Bridgemarsh Island. In his logbook Henry writes that on the Sunday the wind was light and the weather fine. He and Walter took turns to handle the yacht as they sailed up and down the River Crouch. They swam four or five times and spent the afternoon stretched out on the deck before returning to Burnham. The weather was glorious, says Henry, 'with a cloudless sky and a perfect breeze'.[34] The sun blazed all day. They had slept for only four hours out of 26 but were under way for 16 and although the cruise was short, it was, says Henry, very successful.

It was in all likelihood Walter's first encounter with the Cash family; certainly, it is the earliest meeting we can document.[35] The logbook entries for this trip are suffused with a sense of well-being and camaraderie. Henry had sailed before with George Reeve (Walter's brother) but as 'new crew', Walter measured up well.[36] Four years later, when Henry set sail on another trip, he was accompanied by both the Reeve brothers. James Cash was also on board at various times. They left Southampton in the cutter *Diana*, and sailed into the Solent before making their way along the south coast of England. Five days on, they entered Elbury Cove,[37] in Devon, and Henry recorded their impressions:

> We were quite alone and clouds of mist which were slowly melting away in the hot sun alternately hid from view and uncovered little bits of white beach, ruddy cliff and woodland coming down to the water's edge. And the solitary position of the old bathing house with its stone walls green with moss, and red tiled roof and stone steps running down into and under the clear green water distinctly added a further charm to the spot.[38]

The ruins of the bathing house still stand in a corner of the bay, like a little castle, and though its roof has long gone, the stone steps remain. Alongside, a shingle beach falls away steeply. Henry and his companions next headed to Babbacombe Bay and here, too, they found peace and seclusion – as well as an extremely good lunch at a local inn called The Cary Arms. Bright red sandstone

cliffs, typical of the region, provided a vivid backdrop to an unforgettable location and the sailors vowed to return to it whenever they could.

About three days later, near the village of Kingswear, Harold and Grace
joined the *Diana* and, according to Henry, everyone immediately began
'jawing'. As conversations became intense and animated, and his companions
refused to gather for breakfast, Henry's frustration grew. The yacht began lurch-
ing in strong winds and Walter, who had been imploring the others to come
to the table and eat, ended up with a plate of porridge, two lightly boiled eggs
and some bread in his lap. Eventually, he had to change into his pyjamas and
tow his trousers over the side, before hanging them in the rigging to dry. 'We
believe', says Henry, 'that they are egg-stained to this day.'

In this excited, jocular atmosphere, Grace made Walter's acquaintance. She
was 27 and had now been married for nearly three years. On an earlier trip
that she and Harold had taken with Henry, her husband had proved inept as
a sailor, having got the lead line (used to make soundings) in a tangle before
steering the boat off course. This particular vessel, the *Alice Maud*, had been in a
poor state of repair but Henry, by his own admission, was not a patient skipper.
He had cursed his brother-in-law furiously. During the same cruise Harold at
some point had accidentally put his foot through the cabin bulkhead.

The summer cruise of 1899 seems to have been a happier affair. Eventually
the sailors reached Cornwall, dropping anchor along the coast each night to
sleep. Their ideal harbour, the 'perfect anchorage', says Henry, was not just
beautiful; like the perfect life envisaged by the Fellowship, it must offer some
challenge and be difficult to achieve. Then there must be 'snugness and shelter
from any gale'. A quiet cove at Yealm River on the Devonshire coast came nearest to perfection, with its wildness and remoteness:

> We gloried in the scenery, the moonlight, the river, our yacht, our life and
> everything, with the usual ultimate result – namely feeling desperately
> chilly and turning in – thoroughly convinced that life is worth living and
> that real comfort can at length be obtained in a yacht's cabin after a suf-
> ficient struggle with one's blankets.

If community and closeness to nature were important to Fellowship members,
then sailing offered additional pleasures. A guidebook to coastal cruising for
amateurs, written in the 1920s, describes 'the indefinable delight'[39] of over-
coming dangers and difficulties to arrive safely at your destination, without
spending a penny on travel expenses, at places most people would only be
able to reach by ordinary means. The Fellowship of the New Life had been
formally disbanded by now but Nonconformism was a state of mind, almost
an alternative way of being in the world. Henry, for example, was often at
odds with the harbour pilot or local officialdom, always determined to prove

62

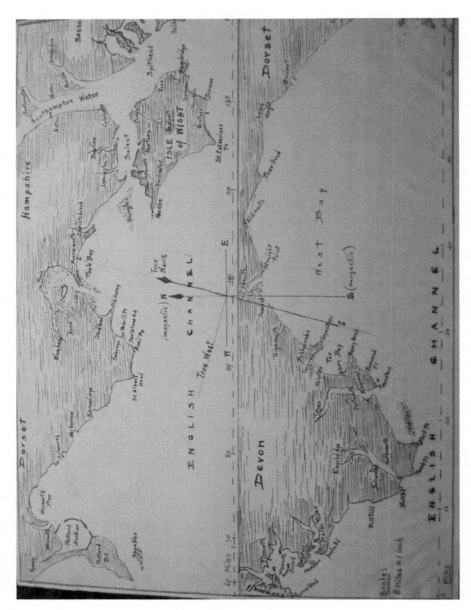

Illustration 5.1 Hand traced map taken from the sailing logs of the Hampshire, Dorset and Devon coastline that Henry and his companions followed in the cutter *Diana* in the summer of 1899

his independence and skill, and to navigate without assistance even through difficult waters.

He and his companions travelled considerable distances at times. Sailing took you completely away from the land, the suburbs, and other people. On board, living was cramped. Everyone, whether male or female, married or not, shared the same quarters and slept on wooden bunks. It would not have been particularly comfortable but there were compensations. In an aquatic wilderness you could build new friendships, see others more truly and confront the elements more directly. It was not without risk, however, and another contemporary manual urges its readers to 'Learn to swim before you go sailing. No mere pleasure is worth risking one's life for, and accidents will happen even to the most skilful sailor.'[40]

The following summer, Henry, Walter, Grace and Harold, and two other young friends, set out again into the flat, slate-coloured waterscape beyond Burnham, this time in a sailing boat Henry had bought, named *Pearl*. Not long after starting the voyage Henry left his companions for a few days to go to London, and Walter took over as skipper (and log keeper.) He had by now begun studying at Guy's Medical School in London and this seems to have been his last trip. The entries in the logbook became light-hearted now, the tone unhurried, sometimes mocking or self-deprecating. Walter was relaxed. Once or twice he seems to have mused about religion, about differences between his background and that of his companions.

One day he remarked sardonically that he had turned out rather late that morning because it was Sunday and he thought 'the day should be marked in some distinctive way'. Walter had experienced an upbringing he would not easily forget at the hands of the Church Missionary Society but the Nonconformist Cash and Oakeshott families, with their interest in a 'new life' and their rejection of social convention, were a challenge to the orthodoxy of such a childhood. Humour was one way to deal with it.

Walter playfully alluded also to Harold's drinking habits. When they had to make a short trip inland for supplies, he wrote that Harold 'being tired? was left in the hedge'. 'On our return', Walter says, with heavy irony, 'we found him in the ditch still more tired, fraternising with a coastguard with whom he was alternately having drinks of tea out of a gallon jar. The coastguard explained that the nearest place for good tea was three miles away and he had to tramp that distance every day to procure it. Such is the passion for teetotal drinks!' Though Walter makes light of his companion's fondness for alcohol, his own evangelical upbringing would have emphasised the 'evils' of drink and the Cash family cannot have been comfortable with Harold's behaviour either. The temperance movement at the time drew its core support from Nonconformist groups, like the Congregationalists, and from the lower middle class. Alcohol was rejected because it caused family breakdown and symbolised lack of control.

When the logbook was in Walter's hands, references to 'the teetotallers' soon became a kind of shorthand for himself and Grace – who was by now attracting his admiration. Usually, members of the crew were given designated roles on board (chief mate, second mate, able-bodied seaman, and so on) but Grace was known simply as 'the Girl'. She clearly had more aptitude for sailing than her husband and on one occasion took the yacht up the river single-handedly. When she swam, Walter commended her floating prowess – and now in the logbook, he began to call her not 'the Girl' but 'the Lady'. Four years her junior and she a respectable married woman, any intimacy between them would initially have been unthinkable. But the teetotallers found themselves alone together on more than one occasion. Further up the coast, beyond Clacton-on-Sea and Hanford Water, when they eventually reached Harwich, Walter wrote:

> WR and the Lady went for a row in the dinghy before turning in. A lovely moon was over Harwich Harbour; it might have been Venice, for the wide harbour seemed to get lost in the sea and the land; and the few houses that could be seen rose apparently out of the water. The craft lay gently rocking – dark specks on a vast expanse of slowly moving silver water – and the slightest of mists hid where the harbour mouth became sea and sky.

As the trip came to an end, and with it the summer of 1900, Walter completed the logbook with a flourish. Photographs of the group now appeared and in an 'illicit digression' Walter created little pen portraits of his companions. Harold ('the curly one') was addicted to conversation, as well as to drink. Henry was a 'sort of automatic machine', addicted to early rising and enthusiasm. Walter, in an obvious allusion to his own fondness for sleeping in, described himself as 'the Lark' or 'Star of the Morning'. But when it came to Grace, he was tongue-tied. Unwilling perhaps to make light of her foibles or weaknesses, or to seem disrespectful, or perhaps to betray his true feelings, the young Walter wrote only 'The Girl or Prayers'. Whatever his thoughts, as the trip came to an end a bond had formed and he was afterwards often seen in the company of both Grace and Harold. The three became friends.

In the early 1900s Walter joined Grace in her walks over the North Downs and across Ranmore Common. By now Walter was in medical school and Grace had joined the Women's Industrial Council. Her investigations into factory conditions for women and girls were well advanced. She was living with Harold (and his mother, Eliza, and his sister, Mary) in Downside Cottage in Coulsdon. Next door, in Fairdene Cottage, was Maurice Adams. Though The Fellowship of the New Life had ceased its formal activities, some of its members continued to live in close proximity to each other for several years. Grace's parents, along with

Illustration 5.2 A photograph taken on the East Coast cruise of August 1900. From left to right, the sailors are: Walter, Harold and Grace

her siblings Jessie, Kate and Henry, had also moved from Thornton Heath and they lived a few doors away in the same road.

When Grace left Harold and her Croydon life behind in 1907, the novel *Attainment* had not yet been published but her decision suggests that Fellowship, brotherhood, perhaps all forms of comradely living, had lost their appeal for her, just as they did for the fictional Rachel Merton. Did 'heaven' prove too hard to find and the vision too difficult to realise? Was Grace frustrated by individual family members and their various foibles? Or by Harold's drinking, perhaps? The answer to all these questions might be 'yes', but the most likely explanation for her actions is that she had fallen in love.

Part II

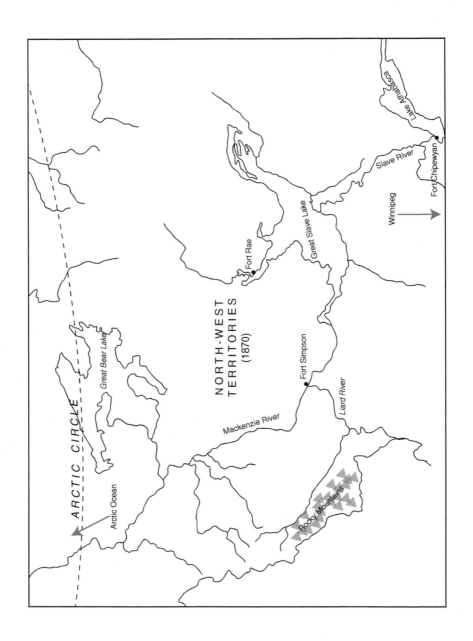

NORTH-WEST TERRITORIES (1870)

ARCTIC CIRCLE

Arctic Ocean

Great Bear Lake

Fort Rae

Fort Simpson

Mackenzie River

Liard River

Rocky Mountains

Great Slave Lake

Slave River

Winnipeg

Fort Chipewyan

Lake Athabasca

6
Answering the Call

Evangelist William Reeve marries Emily Parker and in 1869 they make an arduous journey to the Canadian Arctic to work as missionaries. Life is hard, especially for Emily. Walter is born here in 1876, and he is their fifth child.

A small boy is perched high on a fur-covered box, buggy or sledge of some kind. He wears a Norfolk suit of rough tweed, with a separate, starched collar and bow, knitted woollen stockings and laced boots. It is an adaptation of the sportswear made famous by the Duke of Norfolk in the early nineteenth century. By the 1880s it was popular dress for middle-class English boys.

The child (in Illustration 6.1) is Walter Reeve. He is probably sitting in a photographer's studio and this is a formal occasion. He is three or four years old and though these clothes will keep him warm, they cannot be comfortable. He has been given a whip to hold, perhaps to suggest an outdoor life, or a particular kind of manly future. He wears a serious but slightly bewildered expression.

Walter was born in 1876, a few years after Grace Cash, in circumstances that could hardly have been more different. He was the fifth child of William Day Reeve and his wife Emily, and at the time the family were living in a remote part of what had just become the federal dominion of Canada. The Northwest Territories would see its borders change frequently over the next one hundred years or more, and the Anglican Church would spend many decades seeking to convert the indigenous peoples to its particular version of Christianity. Walter's parents devoted their lives to the cause.

William came from Lincolnshire in England. After a village education, two years of farm work and a training in business, at the age of twenty-two he entered a college for missionaries run by the Church Missionary Society (CMS) in Islington, London (shown in Illustration 6.2). Probably, he had heard one of the many appeals that were being issued in English churches at the time to join the mission movement.

Illustration 6.1   Walter as a small boy

The CMS was (and is still) an evangelical organisation.[1] It was founded in London in 1799, during a struggle between 'high church' Anglican traditional-ists (who believed the clergy should be learned theologians) and evangelicals who were unsympathetic to the Church's exclusive and traditional customs and were calling for a more active, proselytising clergy. Evangelicals thought individual conversion, the Bible and Christ's sacrifice were more important than questions of doctrine, and the movement was at its peak when William began his missionary training.

Even in William's time, evangelicalism had its detractors and those who were ready to identify its dangers and hypocrisies. In 1847, in her novel *Jane Eyre*,

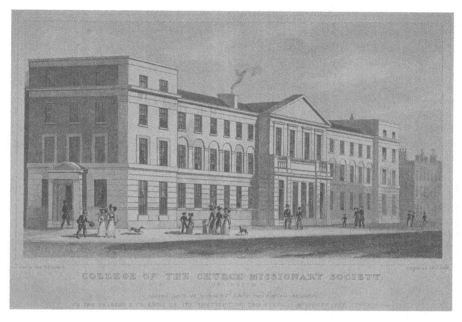

Illustration 6.2 A view in 1827 of The Church Missionary Society College in Islington, London. The college no longer exists but it remained in use until 1917 and nearby St Mary's Church in Upper Street, Islington, continued to serve as a focus for evangelical-ism well into the twentieth century

Charlotte Bronte captures a Victorian child's encounters with the movement when a zealous Mr Brocklehurst addresses an errant young Jane:

'Do you read your Bible?'
'Sometimes.'
'With pleasure? Are you fond of it?'
'I like Revelations, and the Book of Daniel, and Genesis, and Samuel, and a little bit of Exodus, and some parts of Kings and Chronicles, and Job and Jonah.'
'And the Psalms? I hope you like them?'
'No, sir.'
'No? Oh shocking! I have a little boy, younger than you, who knows six Psalms by heart, and when you ask him which he would rather have, a gin-ger bread nut to eat, or a verse of a Psalm to learn, he says "Oh! The verse of a Psalm! Angels sing Psalms," says he. "I wish to be a little angel here below." He then gets two nuts in recompense for his infant piety.'

'Psalms are not interesting,' I remarked.

'That proves you to have a wicked heart, and you must pray to God to change it, to give you a new and clean one, to take away your heart of stone and give you a heart of flesh.'[2]

As a small boy Walter, too, had to learn the Psalms by heart and he was taught by evangelists until he was about sixteen. The founders of the CMS were men who did not question the mission movement or England's imperial activities. Generally, they recruited followers who, like William, came from the lower middle classes and were without a university education. Strength and endurance were necessary and, as Myra Rutherdale explains in her book about women in the Canadian mission field, candidates needed a kind of 'muscular Christianity'.[3] She quotes a Canadian Bishop:

> they must be men of the right kind. Men who are willing to tramp the trail in advance of the train. Men who can find a joy in carrying the Gospel to the lonely settler. Men who with simple reverence can lead in the worship of God a congregation of ten in a neighbour's shack. Men who count it a privilege to be pioneers for Christ and his Church.[4]

Most missionaries did feel important to be representatives of the Church abroad and often, with their transplantation to a new culture, they believed they had achieved greater social status. They left the rigid structures of the English class system behind them and lived for many years in remote regions.

Like most of his peers, William took pride in his vocation. Having completed his missionary training, in April 1869 he married Emily in a parish church in Caistor, a market town in Lincolnshire. Four days later, they set sail for New York. They had to make their journey before winter set in and Arctic travel became impossible. William had had only one month to procure a year's supplies, get married and obtain the necessary outfit for himself and his bride.[5] The CMS preferred its missionaries to be married and last-minute weddings were not unknown. The organisation sometimes commented on the suitability and appropriateness of the match and it always expected the wives to become missionaries themselves. It was assumed you would work just as hard as your husband, but without remuneration, and Emily did so, at some cost to her health.

The journey to what was then called Northwest America took almost five months. The introduction of steamers (like the one shown in Illustration 6.3) would transform travel in the region within the next twenty years but the journey William and Emily now faced was arduous. Once they reached New York they had to go by rail as far as they could, to St Cloud in Minnesota,

Illustration 6.3   This photograph, taken in 1901, shows one of the early steamers in the Northwest Territories, here moored at Fort Simpson, at the confluence of the Mackenzie River (on the left) and the Liard River (on the right). Steamers were introduced in the region in the 1880s, some years after William and Emily Reeve first undertook the journey from England

and then they crossed six hundred miles of prairie to reach Winnipeg. Here, William was ordained as a deacon and the couple were able to rest before setting out again. For the next three months they travelled by open rowing boat, over lakes and rivers, through calm and storm, always accompanied by swarms of hungry insects. The oars in these boats were long and heavy. One contemporary account of the journey,[6] written by fellow missionary William Bompas (who had often made the trip himself) describes how the eight or ten rowers rose from their seats with every stroke. When the wind was fair, the crew spread the sails. When the current was with them, the boat was allowed to drift and everyone could sleep. At times there were 'portages' where the waters were impassable, and the men carried the boats and goods overland.

The Reeves' destination was Fort Simpson. Known originally as the 'Fort of the Forks', the settlement is still today perched high above the junction of the Mackenzie and Liard Rivers. In summer the Mackenzie (Canada's longest river) brings cold, clear water from Great Slave Lake and the Liard brings silt and clay from the Rocky Mountains and beyond. At Fort Simpson the two rib-bons of water meet – one blue and clear, the other brown and muddy – and they run side by side as far as the eye can see, north to the Arctic Ocean. In the nineteenth century the Mackenzie River began to drift with ice from late October, and continued drifting for a month before it set fast. Then it would freeze to a depth of six or eight feet and the spring thaw, when it came, was dramatic. Bompas describes how large masses of ice sometimes blocked the swollen river and were then carried inland by the force of the current. The ice would tear down trees and strip the river banks bare.

It was late summer in 1869 when the Reeve couple arrived. William writes:

> The Fort came in sight about an hour before we reached it. A strong and bitterly cold headwind was blowing which covered us with the spray from the oars and impeded our progress across the river considerably. The river directly opposite the Fort is one mile wide, and a little higher up, where it is joined by the Liard River, it is more than double the distance. As soon as the boats were perceived from the Fort, the flag was hoisted and the people ran down to the landing place to meet them and to give us all a welcome.[7]

Fort Simpson was run by the Hudson's Bay Company (HBC),[8] and William and Emily were first greeted by the Company's staff and their families. Then they were taken to the mission house, just three or four minutes' walk away. This was to be their new home.[9]

And it was handy for them in several respects. By now the HBC's trade with the local Dene peoples was well-established.[10] The Dene brought furs (fox, marten, minx and beaver were keenly sought by Europeans) and exchanged them for guns and household appliances. Since they were on the trad-ers' doorstep, the missionaries had ready access to potential converts from amongst the hunters and their families. They benefitted from the presence of other Europeans, too, since they were able to socialise with them and share medicine and food in times of shortage. When William and Emily arrived there were more than two hundred people living at Fort Simpson, including HBC employees, servants, missionaries and their families, and indigenous peoples; it was, said William, full of bustle and excitement and 'quite like a little village'.[11]

William's first impression of St David's Church was of a comfortable wooden building. This would be the centre of his activities; it stood between the Fort and the mission and was capable of holding one hundred people. The tower and the exterior of the Church were incomplete but five years later, by the time fellow missionaries William and Selina Bompas had arrived, it was finished. Selina described it to her sister in England as 'A pretty, really pretty little church with spire all complete, of wood of course and native built.'[12]

Illustration 6.4   The pulpit in the modern-day St David's Anglican Church, Fort Simpson. The church has been rebuilt several times since its establishment in 1861 but this is the original pulpit and the one William Reeve preached from. The outline of the first St David's church is depicted on its front and sides

William and Emily celebrated their first Arctic Christmas at Fort Simpson in 1869. In his journal William looked back to the year before:

> We little thought then that we should now be so far from all our friends, and in a land where the people and customs are so different. There is nothing here to persuade us that Xmas is at hand, no evergreens, no bells, no meeting of friends, no Xmas cheer in the shop windows.[13]

So Christmas was a low-key affair. Emily had had a stillbirth a few weeks earlier, following a fall on the uneven ground. It had left her frail and there were times when William feared he would lose her. The couple had to make many adjustments to their lives and contacts with those at home were infrequent:

> Of all the hardships which we have to endure in this land of snow and ice (and they are not a few) one of the worst to bear is the hearing from our friends in England only twice a year.[14]

Local customs seemed odd and William struggled to comprehend the Dene world view. Like most of his contemporaries, he regarded non-Christians as 'heathens' and he believed the Dene should be treated like wayward children. His evangelical adherence to the importance of the Sabbath as a day of rest and abstinence was often tested. On one occasion, while he was preaching, some young men began beating the drum and gambling:

> As soon as I had finished the service which I was holding at the time I went to them and telling them of the sin of such proceedings besought them to cease. They did so at once, promised not to commence again, and one of them restored to the owner what he had won. This was very gratifying but as soon as I had finished the service which I took the opportunity of holding with them and had gone to another lodge, they harnessed their dogs and went for a race! What was I to do? I could not go and stop them. I thought it a much less sin to take a drive with the dogs than to be robbing each other by gambling. I therefore took no notice of it.[15]

The Dene were ready to show consideration and respect towards the missionaries, to treat them as they did their own 'shaman', or spiritual leaders. In their minds, there was the possibility that the missionaries might be able to do them harm. The Dene made little distinction between spiritual, economic, political and social worlds. When European medicines failed, as they did in the scarlet fever epidemic of the 1860s, churchmen who sought to comfort and baptise the sick found they were sometimes held responsible for their deaths. The Dene looked for logic and consistency and new ideas were expected to bring results. Reciprocity in personal relations was also important. If they left their child at the mission school, they expected something in return for the privilege of sharing their offspring.

And the Anglicans also found themselves in difficulties through stressing their belief that people were sinful by nature. In 1878 William writes of an old woman who has told him the Indians have no sins; they always walk 'straight' and so do not need to pray for forgiveness.[16] He struggled to convince them that their hearts were naturally bad and without acceptance of this first step in Christian thinking their progress to conversion was severely hampered. Both Roman Catholic and Anglican missionaries challenged the practice of female infanticide and frequently took abandoned or orphaned children of both sexes into their own keeping, often for use as servants. Polygamy was less common than the nineteenth-century missionaries believed but this, too, was opposed by the churches and could be an obstacle to a person's religious conversion.

With the return from England of William Bompas and his new wife, Selina, in 1874, the Anglican mission in the Mackenzie River District received a boost. The couple reached Fort Simpson on a bright, clear and cool day in September, after a moonlit night and what Selina described as an 'exquisite aurora borea-lis'.[17] Bompas had already spent time in the region, learning the indigenous languages and visiting local settlements. He had been consecrated first Bishop of Athabasca (a neighbouring and newly created diocese) a few months earlier but this was probably his first meeting with William Reeve, who was ten years his junior.[18]

On arrival, the Bompas couple moved into the mission house with William and Emily who now had three children. Herbert, Kate and George were doing well and a fourth child, Ethel, was on the way. Walter had not yet arrived but this is the family he would be born into and these, his older siblings. William was initially deferential in his dealings with Bompas, and although they shared a faith and were both dedicated evangelists, they worked in different ways. Bompas was zealous, eccentric and restless, often at odds both with his missionary masters in England and with the native peoples. He lacked respect for the indigenous cultures and actively opposed the CMS policy of enrolling native people as preachers (believing them to be inferior).[19] He travelled extensively throughout the North, preaching, teaching and baptis-ing. William, on the other hand, had his children to provide for and seemed to seek stability. Eventually, despite their differences, William came to like Bompas.

In the 1870s the two families found themselves thrown together and soon they were all facing severe hardship. In the autumn of 1874 there was a seri-ous shortfall in provisions at Fort Simpson and Selina Bompas was concerned that their arrival had brought extra mouths to feed. A sack of flour had already been opened in the boat to feed the men, a bag of rice had got wet and most of it had been thrown away; the biggest grocery box had been left behind by mistake and would not now arrive until the following autumn:

> We have no coffee or cocoa, only a little arrowroot given me by a friend. No corn flour or starch. It is very vexatious but one must cheer up and make the best of it. In the meantime, God is very merciful to us, and sends us so many little helps through the kindness of friends. One has given us 12lbs. of good rice, another a little coffee, another some candles, and one day, to my delight, there came a small keg of butter. We only allow ourselves to bake once a week, using about 5lbs. of flour. This gives William and me a small piece of bread once a day, and all our party, i.e. schoolmaster, catechist and servants – a good sized piece twice a week. Besides this, we have a few biscuits, of which William makes me take one a day. We have a chest of tea and a keg of sugar, so, after all we are fairly off for provisions.[20]

However, the shortages continued and most of the men at the Fort were soon sent away to hunt for themselves. Wives and children were left to live off collected scraps of meat, and before long no food remained. In a letter to her sister, Selina describes what happened next:

> But on the evening of that very day two Indians came in bringing fresh meat. From that moment, the supplies have never failed. As surely as they got low, so surely would sledges appear unexpectedly bringing fresh supplies.[21]

Much as the missionaries may have thought they were there to 'save' the indigenous peoples, at times it was plainly the other way round. More than once the Reeve family faced starvation. Missionaries could obtain meat and fish from the HBC but for a high price and only on condition that they did not trade in fur with the indigenous peoples. If the ice formed unexpectedly early or the caribou took a new route across the barren lands, then everyone, including the Dene themselves, suffered. Every effort was made to store supplies for the winter and the meat could be dried. Selina to her sister again:

> It looks for all the world like a heap of dirty, rough shoe leather. This we have boiled for breakfast, dinner and tea. However, I am thankful to say we are promised a good supply of rabbits through the winter. It is wonderful how one's capacity for food increases in this climate, especially the craving for fat or grease. I used to watch the Indians in our boat with such amaze and disgust – eating a piece of bread with a lump of moose-deer fat like lard! I believe I could now do the same with great satisfaction.[22]

The missionaries and the traders had little or no experience of hunting but William showed willing. The wild ducks evaded him but on one occasion, soon after his arrival, he did succeed in shooting three geese which added 'pleasing variety' to their food.[23] The late 1880s saw some particularly severe winter famines in the region and increasingly outside help was needed. Both the Anglican and Catholic churches in Canada petitioned the government for help and made public appeals for funds. The support that came, however, did not always take an appropriate form. In 1898 Emily wrote to thank the Women's Auxiliary of the Anglican Missionary Society in Canada for their welcome shipment of dry food. Unfortunately, the pearl barley, oatmeal and pepper had become mixed during the voyage and the supply was contaminated; it meant waiting another year for replacement goods. Tactfully, she suggests that to prevent this, the groceries should be sewn up separately in strong cotton bags.[24]

The Northern diet was monotonous, and fresh fruit and vegetables were almost impossible to find. Some of the Europeans planted gardens but the soil was poor. The lettuce was tough and had to be boiled before it could be eaten. Spinach went to seed too rapidly because of the long daylight hours. William planted potatoes and many years later, in New Zealand, Walter would entertain his own children with stories of his early life: 'For breakfast, we had fish and potato, for lunch, potato and fish, and for dinner both of these.'

In December 1874, despite the food shortages of the autumn, Selina was keen to celebrate a traditional English Christmas at the Fort Simpson mission. A few weeks earlier her husband had left for Fort Rae (about two hundred miles away) and although the Reeve family were close at hand, her interactions with them seem to have been limited and formal. By her own account she found herself leading a solitary life. She threw herself into making extensive preparations, including a tree, candles and plum pudding. She invited twelve of the older Dene wives and the meal was laid out for them in the school room:

> My next grand effort was a Christmas tree for the children of the colony. Such a thing had not only not been *seen*, but never *heard* of before, and as whispers of it went abroad the excitement and curiosity were past description.[25]

Selina set her heart on giving a present to every child at the Fort, both European and Indian, and she set to work with what little material she could find. Eventually, there were more than forty presents on the tree:

> Years ago in my childhood, when my busy fingers accomplished things of this kind, my dear mother used to tell me I should one day be the head of a toy shop. How little did she then dream in what way her words would be fulfilled! I actually made a lamb, 'Mackenzie River breed', all horned and woolly, with sparkling black eyes. Also dolls painted and dressed. One infant in a moss bag like the babies here. Some dancing men, moved by strings, one sailor which was my best.[26]

The three Reeve children had, of course, never known an English Christmas and their next would be spent at Fort Rae where their father was sent a few months later to establish a new mission station. It was here, too, that Walter's story would begin.

Fort Rae was a remote HBC post on the northern arm of Great Slave Lake, west of Fort Simpson. Named after the Scotsman John Rae (surgeon, explorer and trader), it sat on a rocky promontory jutting out into Marion Lake. In the 1890s, Frank Russell, a nineteenth-century naturalist, described it as an unattractive spot, 'wind-swept at all seasons'.[27] William arrived in March 1875 but returned to Fort Simpson later in June, when the ice had thawed, to collect Emily and the children. Indigenous people were more numerous at Fort Rae and over the following decades they would be less exposed to European influences than at other posts. William liked living here. He found the locals friendly and, importantly for his young family, food sources were reliable. He worried that with an even shorter summer and frosts from early August, his cultivation of vegetables would cease altogether, but Marion Lake was on the doorstep with an abundant supply of fish, and at certain times the deer passed in their thousands.[28]

There was no Anglican mission house at Fort Rae when he arrived, and no school house or church. The HBC talked about moving the fort to more suitable ground. In the meantime they accommodated William and he hoped suitable premises would be built by the following summer. By June 1876, a month before Walter was born, William was complaining more vehemently than ever to his missionary masters in London. Their entire dwelling was no more than twenty feet square. This included a little kitchen and a porch that had been converted into a store. The only other room, he said, was about fifteen feet square and had to serve for every purpose except cooking – bedroom, sitting room, dining room, nursery, study, schoolroom and a room for the Indians.[29] He does not mention it but it is clear that this is also the room where, on 11 July, Emily gave birth to Walter.

At this time, in remote parts of northern Canada, there were no medical facilities and no professionally trained midwives. Mission hospitals were not established until the 1930s. Sometimes, native women attended a white European mother during childbirth but the two cultures saw the event differently. To Dene and Inuit women, birth was not an illness requiring confinement.[30] Emily had already given birth several times, of course, but she probably felt reluctant to make use of traditional approaches. She had been raised to fear childbirth and she knew of its complications. Also, although European babies were not at risk from the indigenous practice of infanticide, it created anxiety for women in her situation. Writing about the history of the Dene, Kerry Abel observes that female infanticide was practised across the north in times of hardship.[31] Almost certainly Emily knew of cases where female infants had simply disappeared, or been left in the snow to die, because there was not enough food or clothing for them. Male infants were valued more highly since they would grow up to hunt for the group. Fort Rae was a small settlement, and at this time Emily was the only European woman living there. For all her independence and fortitude she must have felt very alone.

William was at hand, of course, and now his midwifery skills were put to the test. Inevitably, white fathers were closely involved in the birth of their children in the North. European practices further south were becoming increasingly medicalised, but here in the Arctic the family could take a central role and celebrate birth as a joyous and significant event. The following year, William is indeed in celebratory mood. In a letter to his sister he made a first attempt at writing in verse – and warned her not to be too severe in her criticisms of it. The would-be bard labours over each of his five children in turn, until he comes to the youngest:

> Walter comes last, a little joy,
> A sweet good-tempered little boy.
> He never cries excepting when
> He's not well pleased, of course he then
> Gives vent as other infants do
> To show that he's an infant too.[32]

What sort of father was William? Was he the sentimental Victorian suggested in the poem or was he the stern and impassioned churchman, treating his children much as he did the indigenous peoples, determined they should understand the Anglican way, learn to fear God and know right from wrong? In fact, most of his children (seven survived into adulthood) saw little of him and family separations were frequent and inevitable. Even the spirited Selina Bompas founders from time to time:

> My loneliness seems very great. I tell myself to work harder and not to brood or despond. I want to live a higher, more spiritual life, and then I should not feel lonely. The early warm days of spring make one feel languid and much depressed.[33]

At Fort Rae William had many frustrations to contend with. Emily's health was giving cause for concern and the Anglican services were not well attended. Whilst admitting that most of the indigenous people here had joined the Catholic Church instead, and were 'nominally Romanists', he insisted that this was mainly due to the intimidating behaviour of the local priest whom they feared. If only he could have more time, he could 'teach them the good and right way'.[34] Bompas was less patient and seemed unwilling to back the endeavour. In his Annual Letter to the CMS in June 1876, William openly expressed irritation with his Bishop:

> Bishop Bompas after giving us the option of removing to Chipewyan or remaining here, has now asked us to return to Fort Simpson and seems to give up the idea of establishing a mission here! My wife's health prevents us removing to Chipewyan this year. Several reasons which I am laying before the Bishop seem to make it inadvisable for us to return to Fort Simpson and I shall be very sorry and disheartened if this place is given up before it has had a fair trial. Some satisfactory arrangement will no doubt be made in the fall when we meet together.[35]

By the summer of 1877, despite his protestations, William had been recalled to Fort Simpson. Here, he continued to struggle with his Roman Catholic adversaries. The Dene played each religious faction against the other, insisting that one gave more tea, sugar or tobacco. Both Anglicans and Roman Catholics believed that the opposing sect was bribing the local people with such gifts. As William was aware, the Roman Catholic missions were better organised and more adequately resourced than the Anglicans. As well, features of Catholicism at this time such as the practice of fasting, the use of the confessional, the charismatic specialist who might lead a person to God and the wide range of saints and spirits who might intervene in daily life were similar to some strongly held Dene beliefs.

Kerry Abel explains that the use by Catholics of rosaries, crucifixes and pictures also resonated with the Dene.[36] These objects were associated with personal power, and in withholding or rejecting them it was feared the Anglicans might be refusing to share their secrets. Though some Anglican beliefs also overlapped with Dene concepts, the ministers tended to be less flexible than the priests and they made greater demands on the Dene before baptism. And there was another important factor at work, in the role of the Metis people who were of mixed European and indigenous heritage. Many had French ancestry and seemed already to have an understanding of the Catholic faith, probably learned from their parents. They made ready allies for the priests.

William persevered against the odds. His evangelising often took him away from his family. At Fort Rae, during his absences, as well as caring for the new baby Walter, Emily busied herself with a school for the Protestant children. She also held the Sunday service and tended those with ailments. When she left, some of the local people wept and told her she was like their mother. A man who had been very sick for many weeks said, 'I shall be miserable, I shall be like a man thrown away when you are gone. No one will be kind to me as you have been.'[37]

But it was obvious by now that Emily was herself very unwell. She was, says William, almost prostrate from a severe attack of rheumatism and he sought permission from the CMS to bring her to England in 1879 for medical treatment.[38] During an earlier bout of pain at Fort Simpson in June 1870, William had feared to leave her alone. On that occasion the pain became so severe that he was seriously alarmed. He carried her to bed and applied hot water to her side to ease her suffering.[39] Probably Emily's hip was affected by what would later be called rheumatoid arthritis. Very likely she experienced some relief of her symptoms with each successive pregnancy when her immune system was altered and hormone levels changed but her pain may well have returned some months after each birth and her mobility could have been badly affected.[40]

It was 1880 when the couple finally returned to England, taking their five small children with them. A sixth, Rachel, was born in England soon afterwards. Herbert, Kate, George, Ethel and Walter were placed in the CMS Children's Home in Islington, London, before their parents (with baby Rachel) returned to the mission in the Northwest Territories later the following year. This time they were posted to Fort Chipewyan, on the western tip of Lake Athabasca, where William was appointed Archdeacon. Emily's health would fail again, however, and records show her residing in England in 1883 and also for a seven year period from 1887 until 1894.

Throughout his time in the Canadian Arctic, William, the evangelist, worked tirelessly and, by his own account, never missed an opportunity to preach the Gospel, to try and draw others to the 'winning side'. He regularly assured his missionary masters in London that he was taking every chance to remind the indigenous peoples of God's mercies, exhorting them to forsake their sins. From the official records a picture emerges of an earnest young Anglican, eager and energetic, striding across open ground, Bible in hand, intent on converting others. Perhaps the Dene were tempted to duck for cover as he approached. Perhaps they were amused or puzzled by him. In any event, they were not the only focus of William's efforts. On New Year's Day in 1870, when he was a dinner guest at Fort Simpson, he reminded his European hosts and fellow guests of his sermon the night before and begged them not to prolong their evening festivities into the morning of the coming Sabbath.[41]

William was spurred on by the intense rivalry between Anglicans and Roman Catholics. The fierce competition for souls smouldered for many years in the Mackenzie River District, and eventually the Anglicans lost. In 1891 William was ordained as the second Bishop of Mackenzie River when Bompas moved to the Yukon, but believing that most indigenous people in the region were at least nominally Roman Catholic, the CMS soon ceased funding the Anglican missions and they were left hovering on the brink of bankruptcy. Unable to access sufficient resources, the missionaries and their families at times faced severe privation. William and Emily were lucky to escape a fire at Fort Simpson in the winter of 1895–6. It destroyed the mission house, with all their winter supplies, personal possessions and records and since, at that time of year, all the water in the region was frozen, the fire had to be left to run its course. And William eventually lost Emily in odd and tragic circumstances in 1906, when at Athabasca Landing he upset their buggy into a puddle and Emily contracted pneumonia. She died of heart failure a short time afterwards.[42] His grief must have been compounded three months later when Bompas, his predecessor and by now his friend and support, also died.

In Canadian history William Reeve tends to be eclipsed by the charismatic and more heroic figure of Bompas but he is perhaps more typical of Victorian missionaries. He was a stolid campaigner, with an unshakeable belief in the worth of the missionary enterprise. He lacked the flair of his better-known contemporaries but he was an indefatigable and determined foot soldier. He found satisfaction in every well-attended church service; his faith was bolstered

Illustration 6.5 William Day Reeve in the 1890s, soon after his consecration as the second Bishop of Mackenzie River

by every conversion to the Anglican way, every child that learnt to read the Bible. In 1892 he recruited a new missionary, a young Canadian minister named Isaac Stringer, who would become a celebrated figure.[43] On his return visits to England, William ensured the publication of the Gospel that Bompas had translated for the indigenous peoples of the region, and he was particularly proud to have ordained the first native minister in 1893, something Bompas had refused to contemplate.[44]

Walter's father remained a significant figure in his life but the two were parted when the boy was small, and his upbringing and education were put into the hands of others. The CMS cared for all the missionaries' children much as if they were orphans. Emily was able to take some of her offspring to live with her for brief periods in England, when she returned there for medical treatment, but the children's contacts with William were infrequent and often several years apart. William married a widow from Toronto a year after Emily's death but he never returned to live in England.

7
Not Much 'Home' About It

In the early 1880s Walter is brought to England and, with his siblings, put into the care of the Anglican Church. Later, he continues his education at Monkton Combe School, near Bath in Somerset.

In her novel *Castle Blair: A Story of Youthful Days*, published in 1878, Flora Shaw writes sympathetically about the plight of Victorian children who were separated from their parents. After being sent from India to live with an uncle in Ireland, her two young protagonists, Murtagh and Winnie, are 'making themselves ill with pining'. While abroad they have suffered from fever and now they have cropped heads, 'wizened yellow faces' and 'little sticks of arms'. They feel abandoned and cannot be persuaded to take an interest in anything. Eventually, they recover enough to spend whole days in the mountains where they go 'rampaging like wild animals'.[1] Eleven-year-old Murtagh is particularly rebellious, angry and impetuous but he is brought to heel in the end when an ill-conceived escapade goes dangerously wrong. At this point it is decided the boy must be sent away to school to learn discipline and common sense.

Orphaned and abandoned children feature widely in Victorian fiction. Sometimes, like Murtagh and Winnie, the youngsters find themselves living with unknown relatives; sometimes, like Walter and his siblings, they are sent to institutions. A school (which was also a home) was needed for the children of Victorian missionaries because their parents often lived in harsh conditions in countries where the climate was considered unhealthy and schools were lacking. In remote Arctic Canada, William and Emily had at times taught their own children. At both Fort Rae and Fort Simpson, the young Reeves had sat alongside other European and indigenous pupils in the mission schools, but undoubtedly the plan was always that, like most missionary children, they would return to England for the bulk of their schooling. So, in 1881 (the year Grace and her sisters enrolled at Croydon High School for Girls), a very young Walter found himself in the CMS Home for Children in Islington, London (Illustration 7.1).

Illustration 7.1   A view of the Home in Highbury Grove, Islington, London. The house no longer exists

He was only four, the youngest age at which he could be admitted. His older siblings (Herbert, Kate, George and Ethel) were with him, and Rachel joined them a few years later when she, too, was barely four years old.

Most of the missionary children had been born overseas and had never seen such architectural grandeur as confronted them now. Many years later a former pupil recalled her first day in the Home:

> My introduction to 1 Highbury Grove is still vivid in my mind. My sister Nelly and myself were small things of five and three quarters and seven. How huge the building looked. I remember proudly reading the verse facing the front door at the end of the entrance hall: 'The children of Thy servant shall continue'.[2]

Islington, thanks largely to the activities of the CMS, was now firmly established as a 'missionary parish' and it had been expanding rapidly.[3] From 1830, horse-drawn omnibuses had allowed clerks and artisans to join merchants and professionals in living further from their employment. The district had now many fashionable squares and substantial, well-built houses, like the one occupied by the CMS in Highbury Grove. By the 1880s poorer people had started to arrive too, displaced from their inner London homes by the building of railway stations and goods yards and, like Hackney (where we first met the Cash family), Islington was losing its appeal for the better off.

At first the Reeve children must have been confused by their new surroundings. One missionary child of the 1850s describes how, when he arrived in England, he was a stranger to the commonest things. His first experience of a railway station, with its noise and smoke, alarmed him and a visit to Regent's Park bewildered him. He did not know a primrose from a buttercup but, he says, if he had seen a banana or an orange tree, he would instantly have felt at home.[4] Walter and his siblings had different memories, of course – of frozen rivers, the charging caribou and the shimmering lights of the aurora borealis. Islington's urban landscape provided a dramatic contrast to the sub-Arctic terrain they had known.

In 1881 there were 35 girls and 36 boys in the CMS Home. New arrivals had passed a medical examination and been pronounced 'free from mental and bodily disease'.[5] They had each brought with them a travelling box containing precise numbers of various articles of clothing, including pinafores, stockings (in black cashmere for the girls, dark merino for the boys), underwear, shoes and boots, brushes and combs, and a Bible. Regulations said that half of these items must be new and the other half in thoroughly good condition.[6] It was the clothing a middle-class child needed for an English climate but the Reeve children had often dressed for extreme cold and some of their fellow pupils had lived in hot countries, like India. There were many adjustments to make and new pupils felt their isolation most of all:

> The first few years of my school life at the old 'Home' at Highbury were not very happy, and I look back with much pity at the somewhat forlorn little person I was then, plunged into the very Spartan life which we led when at school.[7]

To this nine-year-old girl, 'there was not much "Home" about it'. Her words were echoed by an adult Walter when, many years later, he told his daughter in New Zealand that there was 'not much TLC' at the Home. They simply didn't understand children. The little girl quoted above found that the time she was allowed to spend after prayers every evening with her brothers became very precious. Family groups commonly arrived at the Home together and most children were accompanied by several brothers or sisters. As a four-year-old, Walter, too, would have been grateful that his siblings were on hand. Herbert, the eldest Reeve child, had a responsibility in the absence of their parents to keep an eye on the younger ones. Herbert would leave the Home in the mid-1880s and later he taught at a school run by the London Missionary Society in Blackheath, south London. This society had similar aims to the CMS but was run by Nonconformists and the school (known as the School for the Sons of Missionaries) took boys from non-Anglican backgrounds. Herbert was well-liked. He understood what it felt like to be the son of a missionary and the boys remembered him long afterwards for his interest in their activities:

> It was he who helped us with our fretwork and other hobbies, and ordered our wood in bulk so that we got it cheaper; and who, after Lights Out, used to tell us a story, often a biblical one retold in a novel fashion that gave it a new vividness.[8]

Herbert also coached the football team, wrote light pieces for the magazine, played the flute in school entertainments, and on one occasion composed the words for an 'imperialist cantata'. Eventually, Walter also became an all-rounder like his brother.

Both the boys did well academically in the Highbury Home and along with their siblings received prizes for their school work. An eight-year-old Walter was rewarded for his efforts in both Scripture and English. The boys' education was designed to fit them for further study in a public school or (eventually) a university. They took courses in divinity, classics (including Latin and Greek), Mathematics, English subjects (including History and Geography), French and Drawing. The girls at the Home did not study the classics or advanced mathematics (as Grace did) but they did take German and Music.[9] All the children followed a strict routine:

> In those days, we had morning and evening 'prayers' in the big hall – a hymn, reading of the Bible, followed by a long (as it seemed to me) extempore prayer. We knelt in long rows at forms, girls on the right of the centre square table, where the Director was, and boys on the left.[10]

In the GPDSC schools, parents were influential. After all, they were shareholders in the Company and paid fees for their daughters' education. In Croydon in the 1870s, when parents criticised the way the High School was being managed, their complaints were investigated. In the CMS Home, parents were kept firmly at bay. They received 'candid and full reports of the progress and general conduct of their children', but there could be no special pleading.[11] Missionary children were handed over into the care of the Anglican Church and, once admitted to the Home, external contacts (including visits from parents) were strictly limited.[12]

On the first Monday of every month, the children were allowed out for the day. Most of the parents were overseas, of course, and had to rely on friends and relatives to help. The children waited eagerly:

> Those whose memories go back to Highbury recall Sunday evening prayers, when we sang with gusto, 'We expect a bright tomorrow' and the scatteration next morning when lucky folks who got invitations from friends went out for the day.[13]

Some of the visitors were friends of the Church rather than family friends and they took large groups of children to London Zoo, Epping Forest, or even Brighton. Presumably William and Emily saw their children from time to time during the 1880s when they made return visits to England, but although Emily stayed at one point for seven years, her poor health often prevented her from actively caring for them. Even Walter's sister, Ethel (who had severe spinal curvature) had at times to be cared for by friends of the organisation, rather than her mother. All the children were encouraged to write to their parents, but letters could take many months to reach their destination and separations of eight or ten years were common. A missionary child of the 1890s recalled that often he failed to recognise his parents when they did turn up. To him and many of his school fellows, 'the returned parent was a complete stranger'.[14]

During British rule in India, which began in the 1850s, the children of traders, soldiers and civil servants similarly experienced long family separations.

A daughter of the Raj in the 1920s and 1930s later reflected that it was harder for her parents than for her:

> They don't have that intimacy with their grown up children when they come back from India that they might have had. My mother had a mental picture of them retiring and us being there for them but we weren't. They weren't there for us and we weren't there for them.[15]

In the CMS Home the role of the absent parents was taken by the Director and his staff who were often influential and closely involved in the development of their young charges. When Walter first arrived the Director was the Reverend Alfred Shepherd who was remembered by one former pupil as 'a young man of arresting appearance with a very charming smile'.[16] Shepherd kept strict discipline and the children feared punishment and disgrace, despite remembering moments of kindliness and paternalism, as this child clearly did:

> I can remember the first time I had occasion to be sent to him for some childish fault. He was extremely kind and fatherly and I was soon melted to tears. I think what pleased my little heart most was that he asked me if I were called by any pet name at home, and always after that he used that name when speaking to me. I can remember that he had a great but concealed sense of humour.[17]

One Sunday evening Shepherd took this young girl and her friend to hear him preach at a church nearby:[18]

> On our way back we walked with him, and I know he tried to draw us out. My little friend was shy and silent and I did most of the chattering, until he led the talk to more serious things. In reply to something he said I murmured somewhat petulantly, 'Well, we can't be *perfect*.' He said in a tone I have never forgotten, 'Be ye therefore perfect even as your Father in Heaven is perfect.' We turned in at the big gates of the house, and I was silenced but not convinced. Mr Shepherd meant to make me think, and often and often in those far-off days did I wonder and ponder over those words.[19]

Another pupil of the 1880s recalled the doctor at the Home who was 'a sort of Father Christmas':

> Who of those days will ever forget his Malacca cane, poked in through the door of the dining hall at the boys' end during Friday dinners? The legend was that Dr Allan was addicted to 'Friday pudding', a sort of baked suet pudding rich with currants and raisins. Dr Allan with his cheerful greeting and

teasing ways could make any invalid (or malingerer?) feel well on the road to recovery.[20]

Dr Allan had some serious outbreaks of disease to contend with and, during Walter's first year in the Islington Home, there were thirty-six cases:

> It has been a year of illness and therefore, a year of prayer. It has pleased God to send us four months of scarlet fever, two months of measles, and one case of typhoid fever lasting the whole of this last term.[21]

Little Walter was lucky to escape serious illness but he must have been frightened to see so many other children succumb. The typhoid case (a little girl called Amy) was a particular worry to the CMS authorities. The disease was transmitted by contaminated food or water and in 1880 was responsible for over 6,700 deaths in England and Wales (1.3 per cent of all deaths in that year).[22] Its elimination was made more difficult because symptomless carriers could pass the disease to others. Sanitation and hygiene arrangements in the Home were reviewed. A volunteer nurse eventually took Amy out of London to convalesce – and in so doing probably saved her life.

When Walter was eleven years old, the Home moved to Limpsfield, in Surrey. At the foot of the North Downs, about twenty miles from London and 480 feet above sea level, a new Home had been built. The Church authorities believed this would be a better place for children. The climate was healthier and the damp pavements and grimy streets of London would be replaced by crisp, dry, sandy roads. The Annual report for 1886 noted:

> Instead of November fogs, there is found at Limpsfield pure, clean air. Instead of the minute particles of London soot soiling tablecloth and counterpane, the southern breezes come laden with ozone and sometimes perfumed with sea-salt fresh from the Atlantic Ocean.[23]

It was October 1887 when the Highbury children arrived by steam train, probably a 'special'. The railway to nearby Oxted station had opened only three years before. From the station, the children were marched down a dip in the road and up a very muddy country lane to 'a great red building standing up gaunt and bare on a hill top'.[24] An advance party of pupils from Islington had already brought back handfuls of wild flowers from the local common and stories of the grandeur that awaited them all. The new Home (which became known as St Michael's[25]) was built of brick, with hollow walls on the sides exposed to the weather, and window frames of stone. The large, lofty dormitories and schoolrooms had fine panelled walls and the corridors were lined with black and white tiles, so different from the grey slate of Highbury.

The building faced a common that was without trees or shrubs, like a sandy, red desert. The chalk Downs were in the distance, says a former pupil, 'with nothing between us and them but fields with stiles, the red and gold sunsets watched from the corridor windows, and when spring came, the nightingales in the holly trees, and the spring flowers in such lavish profusion'.[26]

A new Director had taken over. The Reverend F. V. Knox was later described by one of Walter's contemporaries as 'a dark replica of Henry VIII'[27] and the children soon came to dread his quick and decisive footsteps:

> They struck a thrill of awe and 'horrid expectation' into the blood of every boy who heard them. Mostly his mission would be trivial; sometimes it would be frankly absurd, for he was undeniably eccentric and whimsical in many of his ideas. But nothing ever dispelled the air of majesty and absolute authority which he carried with him everywhere.[28]

Religion still dominated the children's lives at Limpsfield, and they had now a Chapel of their own. Knox insisted that they chant the psalms in particular ways, every morning and evening, and he took them to hear the choristers at St Paul's Cathedral and Westminster Abbey. By the time the children left, they knew the psalms by heart.

Knox was unpredictable and capricious but always enthusiastic and genuine. His canings were rare and 'notoriously innocuous'.[29] He was remembered as a magnificent teacher and, unusually for a man at this time, he believed in education for girls, too. He taught the older girls and boys together because he believed they could help each other learn. And he extended schooling for girls in the Home, who were now able to stay until they were aged eighteen, instead of having to leave at sixteen.

It was also thanks to Knox that a swimming pool was built. Most of the stone for the foundations of the pool was collected from the grounds, the children themselves gathering 170 yards of it. Local residents helped with donations and at a total cost of £330 the pool opened in 1888. Knox put it to immediate use:

> He was teaching us to swim and float, and one thin boy (now a distinguished lawyer) found the latter feat quite beyond his powers; each time he went back on the water in the floating position he started to struggle and splash instead of lying still. Finally, Knox said that whatever happened he *must* keep rigid. This time he did keep rigid and with fascinated eyes we watched him, stiff as a poker, sink back and back until his head rested on the bottom and soles of his feet emerged, upturned to heaven. In that attitude,

I verily believe he would have remained until now if Knox had not sent us to his rescue.[30]

Walter retained a fondness for swimming. Eventually, in the 1920s, he would help to build another swimming pool, this time in faraway Havelock North, New Zealand. He now had young children of his own and in the warm southern summer, outdoor swimming was popular. Perhaps he thought back to his Limpsfield days as he led the local community in raising funds. In 1921, at the official opening of the Havelock North pool, he was invited to take the first dive, but his children were embarrassed fearing that, as usual, he would bend his knees and land flat on the water.[31]

Walter left the CMS Home in Limpsfield in February 1891, aged nearly fifteen. By the end of the following year all the Reeve children had left. It was fortunate timing because in 1892 diphtheria broke out in the Home and the surrounding district. The Reverend Knox and his wife nursed many of the young patients and years later Mrs Knox could still hardly bear to talk of it.

Children were at particular risk of contagion. The disease caused them to develop hoarse, choking coughs and was known as 'the strangling angel of children'. The whole establishment was broken up and the school thoroughly disinfected. Every crevice was sealed and the building pumped full of sulphur. No sooner had everything been reassembled, however, than the disease broke out again. It was the worst sickness since the Home was first established in 1850 and it must have been the very thing the CMS authorities were hoping to avoid with the move to Limpsfield. Ironically, Knox had to conclude that atmospheric conditions had been conducive to the spread of the disease. The new Home sat high on a bare hillside, exposed to the prevailing winds, and Knox now urged the planting of protective trees and shrubbery.

Diphtheria was carried on the air (and in unsterilized milk) but it remains one of the most obscure of all infectious diseases. In the nineteenth century a link with ground water was suspected, but at the Home, the water and milk supplies, and the drainage and sanitation were all eliminated as possible causes. Close living, of course, helped it to spread. The first nine children who caught diphtheria during the initial outbreak were those who sat and also sang together. With one exception, the only girls who caught it were in the choir. Diphtheria is more fatal to girls than to boys, but at the end of the year Knox did not have to report the loss of a single child. Indeed, the death rate amongst children in the Home compared favourably with other schools at this time. And there were many other cases where lack of sanitation was not to blame for outbreaks of diphtheria. At the turn of the century infant mortality still accounted for almost one death in five and it was some years before improvements in living conditions and childhood nutrition saw a notable drop overall in the number of child deaths.

But Walter was by now out of harm's way. He had earned a leaving scholarship and a few months later was able to continue his studies at Monkton Combe School, near Bath, in the heart of the Somerset countryside. This school had been set up in 1868 to train boys to become missionaries. It adopted the customs and practices of other Victorian boarding schools and, although most boys did not now leave for the mission fields, the school retained a strong evangelical ethos. As a senior pupil, Walter was expected to wear an Eton jacket and collar on Sundays, and a bowler hat. His boots were cleaned by a servant. He was not allowed to read secular books by authors such as Charles Dickens or Walter Scott, and most magazines were forbidden.[32]

In 1891, when Walter arrived, there were 54 boys in the Senior School and by 1893 this figure had risen to 60. In wintertime there were only small coal fires in the classrooms and no heating in the dormitories. Oil lamps were used for lighting and there was very little proper sanitation. Food was plentiful but often poorly cooked. Walter's headmaster was the Reverend R. G. Bryan, a Cambridge-educated vicar. He was a tall, austere figure, with rigid and puritanical views – and he was the last in the string of forceful and eccentric churchmen who populated Walter's schooldays. Bryan was remembered for the way he paced the garden every morning, cloak flying, praying for each boy in turn. He delegated authority to his prefects who held kangaroo courts and punished their fellow pupils for misdemeanours such as smoking, throwing water, swearing, and talking in church. Punishments might include verbal warnings, writing lines, or (most frequently) 'cuts to the person', administered with a cane.[33]

Happily, Walter does not feature in written records of these proceedings. He successfully completed both his Lower and Higher Oxford and Cambridge Certificates at Monkton. As a member of the First XV rugby team, he was 'a very light forward' who distinguished himself in dribbling now and then.[34] Walter also won the three-legged race – and in his first year he obtained a Distinction in Additional Mathematics.

Walter's sporting memories might well have been amongst his happiest. Many of the school's practices were unremarkable for the times but in the 1880s and again in the 1890s, its evangelical roots became strongly apparent. More than once, waves of religious emotion swept the school. Meetings were held where the boys declared their Christian faith, gave witness and illustrated their personal experiences of conversion by reading from the Bible. It was not the first time Walter had lived through a period of religious fervour. He was only five or six when the Director of the CMS Children's Home in Islington gave thanks to God for a spiritual awakening amongst the pupils there.

Illustration 7.2 Monkton Combe's First XV in 1892. Walter is standing on the right at the far end of the middle row, arms folded

No doubt some boys were strengthened by these events and their faith bolstered. In 1895, sixteen per cent of Monktonians were ordained and six per cent became missionaries.[35] It is certain too, however, that other boys reacted angrily and suffered lasting hurt. In the 1850s, George Stallworthy was a pupil at the School for the Sons of Missionaries in Blackheath where Herbert Reeve later taught. Over fifty years afterwards, Stallworthy declared that he had hated the Commandments ever since his school days when he had had to recite them every morning at breakfast – in Latin. He resented the efforts of others to dictate his spiritual beliefs and to bind him to one church or tradition. He recalled that to love the God he had heard and read about was a sheer impossibility:

> It is a psychological fact which I suppose others can match, that a thick murky cloud of wrath blackened for me the whole horizon of my young days, until I touched the verge of manhood.[36]

Walter also became disaffected. We have little information about what he did immediately after leaving school in 1893 but by 1900, when he was sailing off the coast of England with the Cash family, he was openly, if light-heartedly, challenging the religious tenets of his childhood. Herbert, too, later rejected his evangelical upbringing. In 1899 Walter entered medical school in London. Many times, while he was in the care of the Church authorities, he had known other children become ill and even die. As a doctor he would be remembered for his concern for others but the Church played little part in his adult life.

8
Behind Closed Doors

Soon after her marriage to Harold, and about a year before she and Walter meet on board the Diana, *Grace joins the Women's Industrial Council. She quickly becomes one of its most able and respected workers. She now spends long hours in factories, observing and reporting on working conditions for women.*

> For mealtimes there must always be
> An hour before the clock strikes three;
> For tea another half is due
> Before the working day is through;
> And on the shorter Saturday
> There's half an hour for food or play.
> No woman at her work must take
> More than five hours without a break.[1]

These verses appeared in 1900, in a pamphlet published by the WIC. You could purchase the complete poem (called 'The Rhyme of the Factory Acts') in a leaflet (costing one penny) or on ornamental card (for one shilling) for display in factories and workrooms. It sold well and was still in demand five years after its first appearance. In the best tradition of labour chants and songs, the rhyme was meant to remind factory owners of their legal responsibilities but also to empower their employees. The writer was Clementina Black, who had been the driving force behind the establishment of the WIC in 1894, and later became its president.[2]

By 1898 Grace had joined the WIC. She was married and now about twenty-six years old. Initially, the organisation's work was supported by a trust, but funding was withdrawn in 1900 and it came to depend more and more on donations and subscriptions from its members and sympathisers, as well as sales of its publications, like the 'The Rhyme of the Factory Acts'. As well as the donations that she and Harold now made to support the WIC's work, Grace took out a regular subscription. She was not as well placed financially as some of her colleagues, but with Harold in reliable employment she did not need to find paid work herself.[3]

The organisation had grown out of the women's trade union movement. Its members sought to investigate and reform employment for women and, although their work was unpaid, it was not charity or philanthropy in a traditional sense. Unlike the Dorcas society that both Harold's mother and Grace's mother belonged to in Thornton Heath, the WIC had its sights set on fundamental social change. Collections of clothes for the poor might provide local relief in the way a sticking plaster covers a wound, but they would do nothing in the longer term to improve the lives of working-class people. 'The Women's Industrial Council means business', noted one commentator in 1897.[4]

The Factory Acts that Black's poem refers to were passed in the nineteenth century and reflected changes that were occurring in the workplace. The Factory Act of 1891, for example, raised the minimum age at which children could be set to work from ten to eleven years. It also required machinery to be fenced off, and set limits to working hours for women. The WIC monitored breaches of the Factory Acts. Many years afterwards Margaret Bondfield (one of Grace's colleagues in the organisation in the late 1890s), wrote in her autobiography about the strategies the women used to alert the authorities to malpractice. Sometimes, they suggested to a factory inspector that he might stop some of the girls in the evening when they were leaving at the end of their shift and ask them to open their baskets. The baskets would be found full of work that the girls were required to take home and finish. At other times, the women heard about factory girls having to work all night. The girls had been told to knock three times, they said, to be admitted. 'By a curious coincidence', says Bondfield, 'our friend the factory inspector also knocked three knocks that Friday.'[5] His scam having been exposed, the employer was taken to court and fined.

In addition to monitoring breaches of factory legislation, some WIC members were also actively investigating employment opportunities for girls and women in the trades. Grace knew about this investigation work before she joined and she had crossed paths with several of the women involved. Once, at a meeting of the Fellowship of the New Life in 1895, Amie Hicks (who had emigrated to New Zealand in the 1860s but was now back in England) had lectured on the importance of supporting the WIC's work.[6] Hicks had been a close colleague of Black's on the Women's Trade Union Association, the WIC's forerunner. And in 1896 Black herself spoke at a meeting jointly organised by the WIC and the Fellowship. It was essential, she argued, to know the facts about women's employment.

For example, there was a cheap workmen's train, by which girls from the eastern districts of London had to travel to get to their places of work. The problem was that it arrived at Liverpool Street station at 7 a.m. – about two hours before most of the girls began their shift. The interval was spent 'in wandering about the streets'.[7] The WIC proposed a census of the women and girls using that train, so they might ask the railway company to arrange for a later service. A few years later, under continuing pressure from the WIC and several trade unions, the train companies finally agreed to provide cheap trains for workers up to 8 a.m. each day.

Grace was pragmatic and the WIC's methods appealed to her. What was the point of proposing change without a full understanding of the problem? Later, she would become known for her ability to grasp facts, and to 'look all round a question'.[8] Many workers were employed in places with poor sanitation, and they were paid low wages for long hours. 'There were good employers in the metropolis, but there were those whose greed knew no bounds,' said Sir J. Hutton, Chairman of the London County Council (LCC), at the inaugural meeting of the WIC in 1894. To loud cheers he added that if it were 'too herculean a task to wipe away the tears from all eyes',[9] the new organisation might seek at least to alleviate some of the suffering.

Change could be achieved, the WIC believed, by a campaign based on a close examination of working women's conditions, their wages, their training and duties. Members saw themselves as advocates for a fairer society. Many (including Grace) believed, too, in votes for women, but unlike the suffragettes they refrained from direct action. They did not chain themselves to railings or carry out arson attacks. When Emmeline Pankhurst and her daughters were making headlines during an increasingly militant suffrage campaign, WIC members, like Grace, were quietly visiting factories and workrooms, collecting data from wage books, organising questionnaires and interviews, making observations, and talking to employers and workers. They worked persistently and methodically to influence public opinion and government policy with their findings. And they made proposals for legislative reforms. They believed that the disadvantages most working-class women experienced were unlikely to be removed by the right to vote.[10] Important though political enfranchisement was, other changes were needed if poverty was to be abolished. Social investigation (rather than direct action) was to be the tool.

Social investigation was popular with organisations at this time (including the Christian Socialist Union, the Home Office, and the National Union of Women Workers) because of a belief that figures carried more weight than the individual case and that a 'true description' of the poverty in Victorian cities would help counter fear and misinformation about the real circumstances of people's lives. Just as Charles Booth, a ship-owner from Liverpool, had hoped to influence government policy with the survey he began in 1886 (*Inquiry into the Life and Labour of the People of London*) so the WIC aimed to encourage reform in the workplace and in its own words, to 'overcome masculine indifference to women's interest'.[11] Booth's study (which lasted fifteen years and ran to seventeen volumes) would eventually provide support for legal and constitutional changes, such as the introduction of the old age pension and the suggestion of legal minimum wages. His *Maps Descriptive of London Poverty* (first produced in 1889) used seven different colours to show levels of poverty and wealth, street by street.[12]

The WIC investigations were narrower in scope but no less painstaking. They were time-consuming and demanded good levels of literacy. The women wrote detailed reports on the trades and collated figures (on wages, for example). The visits to factories had to be carried out during working hours, and if the data was to be representative, it had to be collected from a wide sample of workplaces. Since literate working-class women on the WIC often found it difficult to spare the time away from their employment, the investigations were inevitably undertaken by middle-class members like Grace, whose financial circumstances were less pressing.

Grace signed up for the WIC's Investigation Committee as soon as she arrived. Hicks was the Chair, Black the Secretary, and its other members included Margaret MacDonald (the wife of Ramsay MacDonald) and Beatrice Webb (already known as a social reformer). Soon, Grace was leading investigations into artificial flower-making, French polishing and several other women's trades, including embroidery and cigar-making. She joined the Education Committee in 1900 and by 1903 she was Honorary Secretary of a new Education and Technical Training Committee, Chair of the Investigation Committee, as well as a member of the Executive. With monthly meetings of all these groups, as well as the investigation work itself, she was fully stretched. It was her busiest year and she would have been away from Downside Cottage for long stretches of time. Her focus in any trade was always on the training and employment possibilities for girls and women and she collected her data using interviews, surveys and close observation. She did not take lodgings with working-class families to see their lives at first hand, as Booth had done, but she spent many hours in factories, talking to workers and employers and she published her reports.

The official journal of the WIC was *The Women's Industrial News* (*WIN*) and for the first time in this story, we hear Grace's voice as she writes fluently and compellingly about poverty, working-class girls, education and factory life. Women had few public outlets at the time and her colleagues also joined in contemporary debates through the pages of the journal, expressed their personal views with passion and canvassed those in authority. Margaret MacDonald worked assiduously alongside Grace and after his wife's death, Ramsay MacDonald recalled how she had carried out her inquiries 'with heart and soul'. It was not, he said, a cold-blooded scientific mission. She used to return home 'laden in mind and spirit' and 'no easy optimism saved her from being seared by what she saw':

> Today, I have met women the latchet of whose shoes I am unworthy to loose, and here I sit in happy comfort, and there they now are toiling, toiling, toiling without hope or brightness.[13]

Grace, too, now came into close contact with working girls and women for the first time in her life. Her reports were extraordinarily detailed, intent as she was both on reaching her middle-class audience and on celebrating the skills of the workers she met. The complex processes of artificial flower-making immediately caught her imagination:

> The first glimpse of girls at work in this delightful trade fills one with enthusiasm for it. The long tables at which the girls sit to work are covered with many-coloured fabrics, which their quick fingers twist and pinch into flowers, which are hung, when finished, on wires running the length of the tables above their heads. The bright materials in their hands, and the rows of flowers above, give a gay and festive appearance to the workrooms.[14]

She visited almost twenty London factories and found that most of the work was carried out by hand. All flowers were made on the same principles but each required a different treatment. Firstly, the male workers would cut the fabric (it might be muslin, sateen, velvet or silk) to the shape required for the corolla of the flower. Grace watched as a pile of material was placed under a heated mould and the mould driven down on to it, so cutting sixteen shapes at once. Male employees dyed and painted the corollas with a brush and spread them on porous paper to dry in ovens before the women took over. Depending on the flower, the women might need to shade the corolla more deeply at the edges than the centre. When it was painted and dried, they pressed it into a cup shape with heated irons.

Next, the flower had to be 'veined' and using heat again, the women now created wrinkles or grooves in the petals. 'Then the flower is ready to be put together', explains Grace carefully. 'The stalk, which is made of wire, is crowned with the seed-vessel and stamens. A touch of paste is put on the bottom of the seed-vessel, and the corolla is then threaded onto the stalk and pressed onto the seed-vessel.'[15] The calyx (the sepals that protect the bud of the flower) was next. Then after adding the leaves, the women covered the stalks. They showed great skill, said Grace, in rolling the green muslin deftly and tightly around the wire. When at last the flower was ready for mounting (as part of a funeral wreath, or as an adornment for a hat or ballgown) the most skilled and best-paid workers took over. 'At first it seems a simple thing to do', Grace said, 'but it is amazing how different a flower appears before it is mounted and after.'[16]

She firmly believed the principles of the work could be taught. Anyone with nimble fingers and a good training could do it:

> But the possibilities of the trade are for those who have a sense of colour and form, which enables them to appreciate the exact depth and warmth of the colour in the shading of the petals and to reproduce the nearest imitation to Nature in the shape of the flower they make, who have the artistic taste which enables them to mount with the greatest possible effect, and who possess that inventive faculty which suggests possible or impossible new flowers, and can adapt old models to new uses.[17]

Despite considerable variation in working conditions, the trade of artificial flower-making offered good employment prospects for girls. Typically, a girl earned 2s. 6d. per week on entering the trade (equivalent to about £11.80 today in terms of its purchasing power). She was paid more as she improved, but much depended on the availability of work. Some firms paid from 15s. to 22s. for competent workers, 9s. to 10s. for the slow and 25s. to 40s. for the exceptional. These rates were good for the period.[18]

However, not all the work was done in factories. The fabric was cut, dyed and moulded there but the women carried out the least skilled and least profitable tasks (the threading and pasting of the corolla onto the stem of the flower) in their homes. Just when we feel Grace might be guilty of romanticising this exacting trade, she explains that homework is 'the darker side of flower-making'.[19] Once home, the women worked long hours in badly-ventilated rooms with poor sanitation. Their children were often employed after school or kept at home to work. Wages were significantly lower than for factory staff. These 'sweated' home workers became the focus of national attention after 1906 but the WIC had expressed concern as early as 1897. 'The child who works at home', they said, 'is losing its educational advantages, either by not going to school, or, if it does go, by not having its full strength available for school work.'[20]

Box-making, like flower-making, was another trade with a large number of home workers. Children, like those in Illustration 8.1, were seldom doing work that would help them earn a living as adults. And it could be dangerous, too. In 1898 the WIC reported that a Mrs Jarvis, whose husband was too ill to work, had been supporting herself and her nine children by making matchboxes at home. The light, wooden boxes were spread over the floor and piled high so the paste could dry but one day soon after Christmas, a single spark from the fire set the room ablaze and the entire family perished.[21]

Illustration 8.1   Working at home: making matchboxes (*c.*1900)

The WIC was split over what to do to help home workers. Black wanted to see trade boards set up that would establish minimum wages but MacDonald proposed a system of licensing for home work. A further disagreement in 1908 about how to handle married women's work eventually provoked resignations, including MacDonald's. She wanted to see increases in male wages so that married women had less need of work; Black advocated improvements to housing and co-operative housekeeping so that married women would be freed to take up employment if they wished.

There were other disagreements, too, some relating to broad principles about the best way to effect change. Working-class women stood to gain from the WIC's activities, as the opportunities for training, skill development and wage enhancement were illustrated in published reports, but working-class leaders like Hicks and Clara James wanted to see women workers encouraged to be active and independent. Undoubtedly, some middle-class members patronised the working-class women, and social investigation, no matter how sympathetic the investigator, did not offer opportunities for empowerment in the workplace. Instead, it turned working-class women into objects of study. So Hicks and James decided to put their energies into organising working girls' clubs. With their own experience of industrial work (Hicks in rope-making and James in the confectionary trade) they were better able than many of their WIC colleagues to understand the lives of factory girls. In the clubs they ran classes in physical drill and gymnastics and also in citizenship (where the girls debated women's suffrage and trade unionism) and social investigation was largely left to middle-class members.[22]

Disputes were distracting and unsettling. According to her biographer, MacDonald's decision to leave the WIC caused her acute personal distress which contributed to her early death.[23] Grace's particular passion was the need for technical training, and she managed to remain focused on this. She was keen to see more women employed in skilled industries and argued that the establishment of a day school to teach artificial flower-making would enhance girls' prospects immeasurably. Flower-making required expertise and you could make good earnings. Further, there was no direct competition with men for the skilled branches of the trade and women were not confined to the more menial tasks, as in many other industries.

But under threat from international competition, London flower-makers were declining in number. French women were thought to have more artistic ability, a finer sense of colour, form and style. Clearer skies and cleaner water in France produced more delicate dyes. French workers specialised more and were more highly skilled than their English counterparts. An English flower-maker was expected to make any flower; a French flower-maker, only roses. One employer compared the work of English and French women. The French woman's rose was perfection; 'the Englishwoman's something she would call a rose, though its shape might resemble a shuttlecock'.[24]

The future success of flower-making would depend on the supply of skilled and inventive workers. Flower-makers in London frequently visited the West End shops to study the new samples from Paris, where all the best flowers were made. One flower-maker said 'her fingers tingled'[25] because she wanted so much to make the flowers that she saw there. In another factory an employee called Aggie also wanted to make new flowers. She suggested novel methods that proved highly effective. After that, Aggie always accompanied her employer on excursions to Regent Street and she became so valuable that she eventually earned 40s. a week.

Workers like Aggie were hard to come by, said Grace, and you could not set too high a value on their work. They might suggest a flower which becomes all the rage, or a way of making a flower which results in a more accurate representation of the real one. For example, they might alter a mould or arrange petals in a new way. However, workers who could innovate were the exception. Most of the girls did their work mechanically and unintelligently, and lacked ambition and interest. 'Tell me where to get girls really keen in their work, showing artistic taste like Frenchwomen, and I can make this trade pay at once. We can't get the workers,' said one employer.[26]

The flower-making trade was struggling. Surely, proper vocational training was the answer. From their research, members of the WIC had concluded that many working-class women lived in poverty because they lacked the skills that would enable them to take up better paid jobs. Yet when training was offered, Grace said, it was frequently refused. Why was this? Investigations revealed that it might be because a girl's parents were unaware of the advantages training could bring, or were simply more inclined to educate their sons than their daughters. And the relationship between paid work and family life was complex. A mother might need her daughter at home for domestic reasons, to care for younger siblings, for example, so that she could go out to work. Girls themselves might see wage earning purely as a stopgap before marriage and could be reluctant to invest effort in training. In any case, few technical classes taught women useful industrial skills. They were often held at the end of the day, when women were already tired and needed at home to cook, wash or mend.

So what sort of training was likely to succeed? How could the demand for skilled female workers be met? Grace argued that once she and her colleagues had identified the job prospects for girls, day trade schools could be established – and eventually, she was instrumental in setting up the first such school for girls in London, specialising in waistcoat-making.[27]

As well as locating employment opportunities, social investigation was useful in other ways. In addition to highlighting shortages of labour, the need for girls' technical education, and the impact of international competition on trades such as artificial flower-making, there were broader reasons to pursue these enquiries. In 1903 Grace wrote that in diagnosing the problems a society seeks to remedy, the women found they had to make a closer examination of the causes:

It is as if there were a closed door between us and the rest of the world; and behind that door the industrial section of the community carry on their arduous work. Occasionally the door is opened a little way, and we get a glimpse of what takes place on the other side, but it is only one aspect or one portion of the workers that is seen. We know little of the conditions of labour, the work performed, the people who perform it; and a book like Booth's 'Life and Labour of the People' came like a revelation; it was the opening, or rather the partial opening of the door. The extraordinary difficulty of throwing it wide open is easily explained. A good deal is due to the observance of what we are pleased to call individual liberty and the rights of personal contract; a good deal more to a fear that we may find things on the other side not so pretty to behold as we should like, and thereby be made uncomfortable; something to the fact that investigation is very difficult work and something that in those trades where conditions are bad, those on the other side are not anxious that the door should be opened.[28]

Sensitive to injustice, Grace declared herself ready to challenge convention to get to the truth. Social investigation could help build 'a correct picture of the conditions of labour' by, for example, establishing the actual workings of current legislation. Did the Factory Acts benefit women or did the restriction of their working hours only handicap them further in their efforts to find employment? Future legislation could take its direction from the findings of the enquiries carried out by the WIC. You could get below the surface of prevailing views and, through finding out about other people's lives, an investigator could learn to sympathise with them in their predicaments. 'If you are an intolerant person', Grace said, 'do a little investigative work.' She believed anyone would understand an employer's methods more when they saw the difficulties brought by competition, or changes in fashion or season, or by the disposition of the workers themselves. And they would understand the employee better when they saw the homes they lived in, the hours they worked and the 'sickeningly mechanical' nature of their tasks.[29]

Investigative work had changed her outlook and she had seen for herself that other lives were possible. There was little to envy about some of them but others were also less socially constrained than hers. The research could be taxing but, she explained, there was 'the fine glow of achievement'[30] that came at the end of a difficult interview successfully carried off, or when permission was obtained to study wage books or some puzzle finally explained. You needed tact and caution. You had to free yourself from any preconceived prejudices, and show absolute honesty in dealing with your findings.

Grace may well have found it hard to keep faith with her own principles at times, as her later encounters with the London makers of surgical instruments and orthopaedic appliances suggest. She describes how these firms were generally unsympathetic to the idea of employing more women for skilled work.[31] One fitter remarked that women were mentally and physically incapable of it. Another found them extremely inaccurate in their measuring, ignorant of human nature and lacking even the elementary knowledge that should have been acquired through observation and experience. Grace, New Woman and social investigator, made a deadpan record of these conversations. She reported that she had visited ten firms, but the prospects for women in this industry were few and employer attitudes unfavourable; accordingly, she made no recommendation with regard to technical education. Her investigation work was encouraging her to focus on things that could be changed.

She also found hostility towards women workers in the French polishing trade.[32] However, in her study of the cigar trade, she did not shrink from apportioning some blame to the women themselves. They were slower, and they lacked interest and incentive. She argued that until women could overcome 'the obstacles which were in their character and their lives' and realise the importance of regarding their industry as a career, an 'aristocracy of male workers' would continue to dominate the trade.[33]

It was a blunt assessment. Grace's activities were taking her into places that, as a married middle-class woman, she would never have visited otherwise and she spoke out confidently.

Despite the lack of remuneration, investigation work brought distinct benefits for women like her, and for the first time she could spread her wings. Public campaigning brought women opportunities for salaried jobs in local government and the civil service. Social investigation developed their expertise in methods of enquiry and, as they pushed for training that would make them more effective, so they opened up new professional careers for themselves. Many were drawn away to make distinctive and successful contributions to public life elsewhere.

The women's individual gain was, however, the WIC's loss. In 1905, when Grace was appointed to a salaried inspector's post at the LCC, although initially the news was greeted with 'acclamation', a few months later the mood had changed:

> While congratulating both the L.C.C. and Mrs. Oakeshott on her appointment as Inspector of Technical Training Classes for women under the London County Council we cannot forget that by that, the Women's Industrial Council has lost a most skilled worker, and the [Education] Committee an able secretary.[34]

In 1914 one official observed wryly that so many of the WIC's active members had left to fulfil public functions, 'that the Council might almost be said to provide a training school for the national service'.[35] The provision of career development for its members was never an official WIC objective but many of the women benefitted from their association with the organisation. Nettie Adler, who wrote a heartfelt tribute to Grace after her presumed death in 1907, took over her work on the London Trade Schools and was still warmly acknowledging her friend's 'pioneer efforts' several years later. She left too, to pursue a career in politics and eventually became one of the first women to sit on the LCC. Another of Grace's close colleagues was Helen Smith who became Lady Superintendent of the Borough Polytechnic. And some individual working-class members also rose to prominence. Bondfield, a shop assistant at the age of fourteen and then a trade unionist, carried out covert investigations for the WIC into the work of shop assistants and eventually became the first woman cabinet minister in Britain.

Overall, the impact of the WIC's activities was considerable. By the time Grace left England in 1907 the women had made systematic enquiries into more than twenty trades and reported on these in the *WIN*. They had investigated fur-pulling, typing, upholstery, domestic service, machining, millinery, fruit-picking and laundry work. Enquiries into a further fourteen trades were underway and by 1913, 118 trades had been investigated. It was an extraordinary achievement and one of which they were justly proud.

Members used their varied networks (including trade unions, suffrage groups and the Fabian Society) to disseminate their findings and present their arguments. They invited influential men onto their committees and liaised with other national and international organisations. As well as publishing reports, they lobbied politicians, wrote legislation and spoke at public conferences. They helped get factory legislation tightened. They organised clubs for working girls and took up individual cases of hardship, acting as a 'poor woman's lawyer'.[36]

Astonishingly, despite its many famous members, its successes and the publicity it enjoyed, the WIC was never referred to in contemporary works. Beatrice Webb, for example, belonged to the WIC but neglected to mention it in the general historical survey she and her husband published in 1894 (and revised in 1920).[37] Even today, there is significantly more scholarly and popular interest in the suffrage movement of the time than in the achievements of this small but influential pressure group. WIC members focused unremittingly on the most downtrodden in Victorian society and all their work was shaped by nineteenth-century understandings of women's roles. There was no attempt to challenge the traditional division of labour in factories (where women usually held subordinate positions) and so they managed to avoid confrontation with male trade unions. They focused on industries with good prospects where women were already employed and sought to improve their earning power; they were pragmatic and worked within contemporary parameters, practising what the historian Joyce Goodman has called 'the politics of the possible'.[38]

Some years later, Ramsay MacDonald described his wife's colleagues: 'A small band of devoted women, united by the bonds of the most sincere personal friendship and by a common interest in improving the lot of women and children', they had 'for a dozen years, been doing work the value of which the country little knows'.[39] He compared Grace to Margaret, who died of blood poisoning from an internal ulcer in 1911, aged only forty-one. Like his wife, Grace was, he said, 'cut off at the very height of her usefulness'.[40] And reflecting on the untimely death in 1900 of yet another WIC member, Edith Hogg, he observed: 'Strangely cruel had been the fate of some of those women, sent, one almost would have thought, by a special Providence to help their fellow creatures.'[41]

Ramsay MacDonald designed a statue in memory of his wife which was unveiled in 1914 and still stands in Lincoln's Inn Fields, in central London, where the couple lived. It is a semi-circular bronze tableau, a three-dimensional frieze, in which the central female character stretches out her arms to encircle a group of cherubic infants. 'Her heart went out in fellowship to her fellow women', says the plaque, 'and in love to the children of the people whom she served as a citizen and helped as a sister.'

Bondfield wrote candidly about her feelings of loss when Margaret MacDonald died. At first she had, she said, been 'too absurdly class-conscious' to understand and appreciate the other woman's qualities.[42] But then that discomfort vanished completely in affection and admiration for her colleague. For her part, MacDonald was inspired by Hicks, and by many of the working women she met during her investigations. Grace also formed friendships with several of the women she worked with. Their life stories suggest that if she had stayed in England, she, too, would have achieved a significant profile, probably in public sector education or local politics. Was she close enough to any of her associates to share the truth with them? Their reaction to her presumed death suggests not and there were those, like Margaret MacDonald, who held deeply conservative views about the way women should behave. A less sentimental appraisal of MacDonald than that written by her husband notes 'a moralistic side'[43] to her character and had she known what her friend Grace was contemplating in 1907, she probably would not have approved at all.

9
Girls in Trades

In the early 1900s, Grace begins campaigning in earnest for girls' technical education. In 1904 she is the driving force behind the establishment of the first London Trade School for Girls, in waistcoat-making. A year later, she is appointed as inspector of women's technical classes at the London County Council.

In a short story written by Katherine Mansfield in 1908, the character Rosabel is employed in a millinery shop in London's West End. At the end of a hard day she takes a crowded bus home. Her feet are wet, her skirts coated with mud and she longs for the good, hot meal she cannot afford. Back home in her dingy bedsit she begins to imagine a better life. In her fantasy she changes places with one of her customers, an insensitive upper-class woman, who has beautiful red hair and a well-dressed young man at her side. Rosabel pictures herself in a sumptuously decorated bedroom. The woman's companion becomes her lover, and the luxuries and sensations of privilege become hers, too:

> She would sit down before the mirror and the little French maid would fasten her hat and find her a thin, fine veil, and another pair of white suede gloves – a button had come off the gloves she had worn that morning. She had scented her furs and gloves and handkerchief, taken a big muff and run downstairs. The butler opened the door, Harry was waiting, they drove away together ... *That* was life, thought Rosabel![1]

Since the retail trade had expanded and was no longer the preserve of men, middle-class women (like Rosabel) might find employment as shop workers. Retail jobs had more status than factory work but were no less arduous since you could be on your feet for seventy-five hours a week. Shop assistants in London's West End worked shorter hours than their sisters in the poorer East End, but, as Grace's colleague Margaret Bondfield observed in her enquiry for the WIC,[2] although West End customers could be charming, they could also be rude. They regarded shop assistants as lackeys, expecting them to wait on them as their domestic servants did.

If Rosabel dreamed of a different life, so too did her creator. Mansfield, the daughter of a banker, had fled the conventions of family and country and arrived in London from her native New Zealand in the early 1900s. Determined to become a writer, she soon found herself on bustling streets where all the accents of the empire could be heard. London, the imperial metropolis, was the largest city of the known world. By now it was the world's busiest port and its richest financial centre. The invention of electricity, and with it the installation of interior lighting and lifts, had brought a huge increase in the number of office workers. Every day, lawyers, bankers, insurance agents, stockbrokers, importers and exporters poured into the city from surrounding districts. Politicians and businessmen came too, as well as engineers, accountants, architects, and thousands of clerks.

When Grace began her voluntary work for the WIC in the late 1890s, she, too, travelled regularly to the capital. She took a train from what was then called Coulsdon and Cane Hill station which was within easy walking distance of Downside Cottage. There were five or six rush hour trains every day and a growing army of suburban commuters now left domestic lives behind to travel into London. The carriages pulled by the steam-powered locomotive had separate compartments but no corridors. Middle-class women like Grace travelled in second class, and the journey time to London Bridge (where she had to change trains for Charing Cross) was about half an hour, roughly the same as it is today.

From Charing Cross station it was then a short walk for Grace to the WIC's offices off the Strand. Trafalgar Square, one of the city's most striking monuments to imperialism, was close by and Nelson's massive column just a glance away. London's public art and architecture were meant to create an understanding, an atmosphere celebrating British heroism on the battlefield, British wealth, and the country's sovereignty over foreign lands. But London was a city of extremes and in the 1880s, in Trafalgar Square, amongst the fountains and the pigeons, about four hundred men and women slept rough. Alongside the affluent, often underneath their noses, lived the poor, many on the street or huddled in doorways and under bridges. Grace and her colleagues knew from their trade investigations that these extremes existed and that the capital's children were amongst the most vulnerable. As nineteenth-century London grew, so the numbers of poor increased; abject and degraded, often living in filth and misery, they made up 'almost a city within a city'.[3]

Education might have offered a way out for some but although school attendance had been made compulsory in 1880, many children from the poorest families did not go to school. Instead, they worked, for example, as street sellers or domestic servants. And a significant number of those who did attend school in London at the turn of the century were habitually hungry. The first elementary school teacher to win a government post, Dr T. J. Macnamara, estimated in 1904 that about twenty per cent of working-class children were in a 'hopeless condition' with regard to food, housing and clothing. He describes seeing children in his classroom in winter suddenly seized with vomiting: 'This is not so much caused by the fact that the stomach is upset as that it has revolted against the effect of cold on its empty condition.'[4]

When they left school, most working-class children were employed in unskilled jobs. They might find themselves in factories, labelling or packing sweets, in work that offered little training and no advancement. Grace might have been entranced by the colourful workshops where young women made artificial flowers, but from her other investigations she knew that many factories were unhygienic, dangerous and oppressive places. Margaret Harkness (a radical journalist, cousin of Beatrice Webb and friend of Eleanor Marx) describes a contemporary sweet factory in her 'slum novel', *In Darkest London*.[5] The genre of slum fiction was popular and was intended to shock its readers by exposing squalor and poverty. Harkness took a room in the East End of London, in the poorest part of Whitechapel, and she based her characters on people she met there. In her fictional sweet factory small girls could be found dipping white sugar mice into vats of chocolate:

> Girls brought the mice on trays from the kitchen, and gave them to 'hands' who were mere children. These puny things stood by the vats, inhaling the sickly smell of the chocolate, dipping a finger into a vat and afterwards sucking it. They looked up when the labour-mistress opened the door, then continued their work of dipping and ornamenting, drying and packing, as though they had been bits of machinery.[6]

Most of the factory girls in Harkness' novel had come straight from elementary school and some were as young as ten. By 1900 the school leaving age had been raised to twelve but even after that, lack of enforcement and the need for money meant that poorer children could still be kept at home to care for siblings, sent out to work as domestic servants or employed as unskilled labourers in factories. Better off working-class families, seeking improved prospects for their children, might also face difficulties. For girls particularly, there were few openings. When Grace's colleagues on the WIC visited the homes of 311 girls who had left school, they found no evidence of effective planning or preparation for the girls to take up a skilled trade.[7] A tiny minority might be fortunate enough to win scholarships to local secondary schools where they could join their middle-class peers but for most working-class children, once their elementary education came to an end, opportunities were restricted.

The growth of the empire was creating new markets, however, and the WIC's investigators had found that some skilled work was available for girls, particularly in the needle trades. Soldiers and sailors needed uniforms, wealthy importers and exporters wanted suits, and their wives evening gowns and fashionable hats (like those sold every day in a West End millinery shop by the fictional Rosabel). Clothing was the single largest industry in London and it accounted for almost a third of the city's manufacturing workers but the industry relied on 'sweated' labour, or those on low wages working in poor conditions or at home. Some factory employers were keen to take on girls and could offer them well-paid, skilled work but usually complained that they lacked motivation and training.

Then, in 1901, when she was Honorary Secretary of the WIC's Technical Training Committee, Grace was asked by the progressive LCC to lead an enquiry into ways of improving education for women workers. She was barely thirty but already known as a skilled investigator. Now began the round of meetings, letters and official reports that would eventually culminate in the establishment of the first Trade School for Girls in London. It could be dull, routine work at times – it was a very unglamorous form of activism – but Grace and her colleagues were painstaking and their findings were trustworthy. They knew how to convince the establishment. The LCC was the largest municipal authority in the world and it had been elected in 1889, following the abolition of the corrupt Metropolitan Board of Works (an appointed body that had earned the nickname the 'Board of Perks').

In response to the LCC request, the women visited London polytechnics and studied their syllabuses for technical classes. There were some day classes in 'domestic economy' for girls (including cookery, laundry and housework) but little evidence that these offered training that was of practical use. There were evening classes, too, but these did not usually attract the industrial worker. They were more likely to be taken up by young married women who wanted to learn to cook for their husbands. Grace concluded that girls did not want 'to attend an evening class at the end of a day's work for the purpose of continuing a trade at which they have already spent nine or ten hours'.[8]

Another problem was that employers were not always aware that the technical classes existed. So, in October 1901 Grace wrote to a number of local dressmaking firms. There were evening classes available, she said, where their workers could learn new skills and become more efficient. She urged the employers to arrange for some of their female workers to leave earlier one night a week so they could attend the classes conveniently.[9]

It was a practical suggestion but Grace recognised more subtle barriers to change, too. The evening classes were not a success because factory girls were too tired at the end of the day, but also because they lacked any real incentive to attend them. These girls expected to remain in the workplace only for a short period; in most cases they were marking time before marriage and had little interest in training. They lacked ambition, tended to accept lower wages than their male counterparts and were happy to perform the most unskilled work. Grace called this 'the industrial attitude of women', and although she understood the reasons for it she believed it would have to be overcome if women were to progress. Technical training for girls had to be made as accessible as possible; it should be free, take place in the daytime, and in some cases, maintenance scholarships should be provided. A thorough training would encourage 'the formation of business habits' and discourage 'the present aimless flitting'.[10] The girls would develop an attachment to their work; they would become stronger and more confident as they gained new skills.

Where were London girls going to acquire such training? Existing technical classes were not suitable. There was an apprenticeship system but the WIC had long been critical of it. As apprentices, girls on a low wage were attached to the factory workroom and simply tried to pick up what knowledge they could from other workers. They were not able to learn the job in any depth and often they did not receive an all-round training. There were exceptions, such as a well-known dressmaking firm (Messrs Debenham and Freebody), which allowed its apprentices two afternoons off a week to attend classes at technical institutes, but in general the apprenticeship system was disliked by WIC members and considered to have failed its young participants. Grace's view was that proper technical education should take its place.

In 1904 she spoke at a conference organised by the National Union of Women Workers: 'The education of middle-class girls is safely established in England, and will now keep pace with that of boys', she said. 'In elementary schools the progress of the education of boys and girls goes hand-in-hand. But technical education is a matter in which we appear to be behind other nations for both boys and girls; and the technical education of our girls is not up to the level of that of our boys.'[11]

Opportunities for technical education, like those Grace was envisaging, could only be taken up by parents who could afford to keep their daughters in education a little longer. Even if they could see the long-term value of a day-time technical course in a trade with good employment prospects, they might not be able to forego their daughter's earnings, however meagre. So Grace and her colleagues set their sights on the better off members of the working class (a group she calls 'industrious artisans') and on the lower middle classes too, where parents were more likely to have the resources they needed. If the WIC were successful, they might manage eventually to lift living standards and improve life chances, at least for some.

Grace now put her energies into planning the new technical schools. During a trip to Paris with her colleagues Adler and MacDonald, she visited the French technical schools for girls, or the Ecoles Professionelles, as they were known.

She observed classes and spoke to the women running the schools about their funding, curriculum and achievements. One technical school in the Rue Fondary, a poor quarter of Paris, had opened eleven years earlier with six pupils; by the early 1900s there were more than three hundred on the roll.

The curriculum varied according to the girls' backgrounds and the needs of the locality. At the Rue Fondary school, for example, girls learnt artificial flower-making because local employment in the trade was readily available. The director of another school in the Rue de Poitou, Mme Delaunay, grey-haired and practical, said that twenty years of experience had convinced her that the French technical schools were wanted and appreciated by working people. Some of her pupils had become world-famous dressmakers and milliners – but every girl was required to continue her general education as well as learning her trade. When she left, she was equipped with more than a manual skill. She could take her place in the world and hold it. 'Why? She is educated', Grace explained. 'Clumsy, unintelligent, lopsided ways of doing work can be educated out of any girl. Our English girls, too, might be taught their trades from the brain downwards. Why not?'[12]

The French schools provided the model and inspiration for the London Trade Schools for Girls and Grace used the information she gleaned in Paris to persuade colleagues to back her cause. In February 1904 Helen Smith invited her to a meeting at the Borough Polytechnic Institute to discuss the establishment there of a first Trade School. Grace explained to the gathering why a school in waistcoat-making was especially suitable. This trade was not, she said, subject to seasonal fluctuations; it paid well and skilled workers were in short supply. Further, a waistcoat was made by one trained worker from start to finish so a girl could acquire all-round expertise. At a later meeting Grace made suggestions about the timetable. The pupil should spend the bulk of her time, about twenty-two hours per week, learning how to make a waistcoat. Other than that, some general education (including English, hygiene and domestic science), physical exercise and drawing were to be included.[13]

At first, the course was planned to take only one year but it quickly became apparent that two were needed. It was still shorter than the training offered by the Ecoles Professionelles, but as in the French schools pupils were required to pass an entrance test and to continue with their general education. They had to have their parents' consent and a letter of recommendation from their headmistress. And, as a rule, you had to be aged fourteen. Many girls left elementary school at the age of twelve (or before) and so places on the new training programme were taken up predominantly by daughters of Grace's 'artisan class', or those whose families could afford to keep them in education.

The LCC made a grant of £120 towards initial expenses and on 3 October 1904, with eleven pupils, the experimental school at Borough Polytechnic Institute opened. In January 1905, following letters to local schools and advertisements in the Sunday newspapers, eleven more pupils were admitted. None of the pupils was charged a fee and all were on probation for the first three months. Soon the LCC was making maintenance awards to those in most need, but even then some pupils were forced to withdraw because of poverty. Fee-paying pupils from better off families were accepted from the following year (on payment of 10s. per term) provided they met the entry requirements.[14]

As in the French schools, an advisory committee met regularly and monitored the girls' progress. Most of its members were master tailors and employers in the waistcoat-making trade but Grace was also invited to join the committee in its first year. As well as helping to place the pupils in jobs at the end of their training, the committee helped publicise the course and inspected classes to ensure teachers were competent.

Early in 1905, following one such inspection, the committee noted that the girls' general progress was very satisfactory, but they were struggling with buttonholes which were always too tightly worked, making them hard and inflexible. The committee believed the girls should do less work more slowly and complete more samples; if they made finished waistcoats too soon, there was a danger that they and their parents might think they were fully trained

Illustration 9.1   A class in waistcoat-making at Borough Polytechnic Institute, in the early 1900s

when they were not, and they might leave to take inferior tailoring work. The parents had to be in sympathy with the venture to ensure the school remained viable and the committee regularly invited them to come and talk about their daughters' work and prospects in the trade.

Another early challenge for the committee was to find an expert from the tailoring trade to teach the girls waistcoat-making. The future of the school depended on finding the right trade teacher, but Borough Polytechnic Institute said they would only be able to offer a temporary post which might be insufficient to attract a successful woman away from secure and well-paid employment in the industry. Grace suggested that the person appointed might be allowed to bring her own work into the classroom and have the girls help with it. It was an unorthodox solution but the Education Committee eventually agreed. It would be an interim arrangement for the first year only; the school was an innovation and its progress would need to be carefully watched.

Grace's pragmatism was instrumental in getting the new school underway. The authorities trusted her and a suitable teacher was soon found through personal contact. Though learning how to make waistcoats occupied most of the girls' time, the general education and precision-drawing classes were important too, and carefully linked with the trade instruction. Soon, as well as the intricacies of buttonholing, in their first year the girls were learning about the industrial revolution, the condition of British workers in the eighteenth and nineteenth centuries, and the history of trade. In their literature course selected texts included *A Tale of Two Cities* by Charles Dickens, *Shirley* by Charlotte Brontë and *Silas Marner* by George Eliot. Shakespeare was on the syllabus, too (with *Julius Caesar* and *The Tempest*) and the girls also had to learn some poetry by heart.[15]

In these ways, the general education syllabus was similar to the one a middle-class Grace and her sisters had studied some twenty years earlier at Croydon High School for Girls. English secondary education was expanding in the early 1900s and technical education had its critics. It might enable you to become a foreman or supervisor but it would not help you reach the ranks of the white-collared workers. Some trade unionists believed that secondary schools were the best way to provide social mobility. Probably Grace was mindful of these contemporary criticisms. By including some general education in the Trade School syllabus, she offered those pupils at least a taste of some of the more academic courses that were available to their peers in the secondary schools, even though (unlike their secondary school counterparts) they could not take examinations or obtain qualifications in these subjects.

The young Trade School pupils might have been taught in elementary school to believe that education was not for them but Ramsay MacDonald's memoir of his wife, Margaret, explains that although it was hard and discouraging work at first, this handful of WIC women knew what they were doing. Grace and her colleagues were not merely asserting the right of women, he said, to share the educational advantages of men, 'they were striving to create a new, enlightened, active, responsible womanhood, enjoying the pleasures of initiative, independence and self-power'.[16]

Worthy though the women's aspirations may have been, the girls themselves often struggled. Poor diet and lack of exercise had made some of them vulnerable to disabilities, such as spinal curvature. By 1906–7, when additional trade classes for girls had opened at Borough (in dress-making and upholstery), out of a total of 67 pupils, there were eighteen with left convex curvature of the spine, three with curvature to the right, and two with double curvature.[17] Physical training was needed to counter the effects of long hours of trade work in a stooping position. In the Stanley Gymnasium for Women (named after Maude Stanley, a campaigner for women's education and shown in Illustration 9.2) the girls exercised for about an hour and a half a week, under the supervision of a specialist teacher, to improve their physical development and remedy their posture.

The first Trade School for Girls at London's Borough Polytechnic Institute was a success. Before long, daytime trade classes for girls had opened in various

Illustration 9.2   The Stanley Gymnasium at Borough Polytechnic Institute was opened in 1904, the year the first Trade School for Girls began

parts of London. By 1911, about 638 girls were receiving industrial training, and by 1915 there were eight Trade Schools for Girls in London.[18] There was a demand for similar provision in other industrial centres and the WIC had received enquiries from Bradford, Brighton and Cambridge. Local conditions varied, however, and London, with its wealthy and fashion-conscious upper classes, a willingness to innovate and a strong sense of its own importance, was uniquely placed to satisfy the employment needs that the WIC had identified. The trade school at Borough served as a model and others soon adopted the same entry requirements and a similar curriculum. The Bloomsbury Trade School was the first to have its own premises, rather than being housed within an existing technical institution and it was the first, too, to offer a subject not traditionally associated with women's work. Commercial photography was growing rapidly in the early twentieth century and Grace discovered that rates of pay for photographic assistants were high. Girls were recruited to learn how to develop negatives, make enlargements, remove defects, enhance images and mount finished prints. They mastered delicate and exacting procedures and on completion of their course, like other trade school pupils, they knew that employment awaited them.[19]

Rates of pay for girls leaving the trade schools were good, for their time. The first girls to leave the waistcoat-making school, for example, depending on their competence earned between 7s. and 15s. a week (equivalent to about £33 to £70 today in terms of its purchasing power). Exhibitions of your work brought you to the notice of local employers and your training gave you a distinct advantage in the market place. You could continue to receive support in later years, too, and many girls returned for regular meetings with their former tutors and fellow ex-pupils.

The establishment of the first Trade School for Girls in London was a breakthrough for Grace. The number of available places remained low and secondary schools far outgrew trade schools in later decades[20] but for the individual girls who passed through them, the LCC trade schools represented a life-changing opportunity.[21] In 1910 they were described as one of the LCC's 'happiest inspirations'.[22] Though Grace had disappeared by this time, her colleagues continued to remember her pioneering work and acknowledge her leadership. It was her foresight, zeal, judgement and enthusiasm, they said, that had ensured girls would now have the same kinds of opportunities for technical training as boys.[23]

Others might have rested here, content with this achievement. Like many of her married colleagues, Grace might simply have chosen to continue her public activities in a voluntary capacity, as part of a national pressure group campaigning for changes to working conditions for women. In March 1905, however, she applied for a salaried post as an inspector of women's technical classes at the LCC – and was successful. Few women were employed by the LCC at the time, and the vast majority in white-collar jobs were single. There were 87 applications for Grace's post and ten candidates were interviewed. On a shortlist of three, the other two were single women.[24]

Grace's work now took her away from Downside Cottage for significant periods. In the early 1900s the offices of the LCC were in Spring Gardens, near Trafalgar Square, in a building that has since been demolished. From Charing Cross station Grace now turned left and walked the other way, away from the WIC, and across Trafalgar Square, just in front of Nelson's Column. It was exceptional that, as a married woman, she should hold a salaried post at the LCC, and her appointment was remarkable in another way, too, because most educated women who were employed in the organisation were in jobs that did not stretch or test their abilities.

Philippa Fawcett (daughter of Millicent Fawcett, a leader of the women's suffrage movement) was another unusual appointment and she and Grace would have known each other. Like Grace, Fawcett was able to use her talents in a senior role, as an effective and enthusiastic administrator, and she became a prominent activist on the staff. The social climate at the LCC was one in which radical intellectuals could feel at home. Before the war they could be found in small groups, arguing quietly about progressive ideas such as equality between men and women. Many activists were curious about the lives of working people and preoccupied with finding a cure for poverty. Intellectually, this environment suited Grace. All the same, she may have missed the companionship of her female colleagues and friends at the WIC and she was reminded of her status in another way, too. In contrast to two male inspectors appointed at the same time, who were each paid a starting salary of £400 a year, Grace received only £300.[25]

Still, it was a timely move. At the end of the nineteenth century, Britain was falling behind its industrial competitors and technical training was being neglected. The LCC's Technical Education Board (TEB) set up in 1893, with Sidney Webb as its chairman, had been using charitable money and revenue from a tax on alcohol (known as the 'whisky money') to fund a vigorous expansion of the capital's training provision. The Board had created a scholarship ladder to enable selected children to move beyond elementary school into secondary, technical or other forms of continuing education. During the ten years of the Board's existence, the number of polytechnics and technical institutes in London had grown from eight to twenty-six and day classes had trebled. Hundreds of scholarships were provided for children undertaking technical instruction.

By the time Grace was appointed to the LCC in 1905, the TEB had been replaced by the Higher Education & Scholarships Subcommittee. Webb was still in the chair and he now approved scholarships for the girls who had been selected for the Trade Schools. Both he and Ramsay MacDonald had been members of a Special Subcommittee on Technical Instruction for Women in 1902 and were still actively supporting the development of technical education in the city. Both men were known to Grace through their wives who were members of the WIC. And the LCC had already received several letters and reports about women's technical training from Grace, on behalf of the WIC, so that her work was also known to them. She was perfectly placed to lead developments and she had, in effect, created her own job.

Grace remained an active member of the WIC Education and Technical Training and Investigation Committees and her former colleagues worked with her in her new role. When she was presumed dead two years later in 1907, they grieved for her as if she had never left. That same year, a political shift took place at the LCC. Despite an overwhelming majority in the 1904 election, the Progressive Party (allied with the Liberal Party in national politics) was crushed and the Moderates (known as the Municipal Reform Party) who were associated with the Conservatives, took power. It would have been discouraging to many Londoners with a radical or progressive outlook but it did not mean an end to reform. The Trade Schools for Girls continued to grow in number well into the 1930s, long after Grace's departure.[26] When, in 1907, the WIC received news of Grace's drowning and placed on record their deep sense of loss, they could not have known the truth. Did Grace herself ever see their notice? If she did, we can speculate about how she might have felt – but her colleagues' feelings were unambiguous.

10
Medical Men

Walter's story resumes when he enters Guy's Medical School in London in 1899, aged 23. Here, he is taught by several eminent men, including the controversial surgeon, William Arbuthnot Lane. At this time, medicine is rapidly becoming more scientific and the profession is gaining power. In 1907, his training completed, Walter applies to join the Colonial Medical Service.

> The First Year's Man is the most important man in the hospital. At least, he thinks himself so and we are bound to believe him. He is an interesting study, he is so full of good resolutions, and people with good resolutions are always interesting.[1]

In their satirical magazine, *Guyoscope*, students at Guy's Hospital took it upon themselves to welcome newcomers in this way. Walter became a 'First Year's Man' when he embarked on his medical training at Guy's in September 1899 and, aged 23, he was several years older than his fellow first years. He had left Monkton Combe School in 1893 and we have only a hazy picture of how he spent the intervening years. He is barely visible in official records. He attended a London dinner for Old Monktonians the year after leaving the school and somewhere (probably at evening classes in Croydon) he acquired an additional qualification in Physics. And he had by now met the Cash family, with their radical, Nonconformist passions, and had accompanied them on several sailing trips. His 'good resolutions' had had ample time to incubate.

Academically, however, he was underqualified for medical school. In the late nineteenth century, British students without university degrees normally had to take examinations in Greek, Latin, English and Mathematics before beginning their medical training. At Monkton Combe, Walter had passed both the Lower and Higher Oxford and Cambridge Certificates, and obtained a Distinction in Additional Mathematics. These certificates, together with the extra one in Physics, are noted on his record at Guy's but he had no degree and there is no mention of entry examinations.[2]

Still, Walter was well-connected and middle class. In the nineteenth century medicine had less status than either the church or the legal profession. It stood on the periphery of respectable trades and crafts and it did not attract many from the upper ranks of society. Medical men (and they were nearly all men) often had fathers who were already surgeons or physicians but a significant number, like Walter, came from clerical families. London's medical schools, like the city itself, saw their role in both national and imperial terms. In the early 1900s they were actively recruiting students from the colonies, including America, Australia and New Zealand; an application to study at Guy's, from the son of a missionary who was by now a Bishop, was bound to be positively regarded. Walter must have considered the possibility of becoming a missionary himself, as he was growing up. He may even have felt destined for it.

After all, his school at Monkton Combe had originally been founded to train boys as missionaries and he had received an evangelical education. But, while in the care of the church, he had also witnessed severe outbreaks of diphtheria, typhoid and scarlet fever and he had seen other small children suffer horribly. Experiences of childhood illnesses like these, in Victorian times, could often prompt a young man to take up a medical career.

An informal interest in science might also be a spur to those with medical ambitions. Like many of his contemporaries Walter was curious about the natural world. During his first sailing trip with Henry Cash in the summer of 1895, several years before he began his medical studies, on a moonless night on the River Crouch in Essex, he and Henry went for a swim. As they dived, they were captivated by what they saw. A heatless light, generated by thousands of tiny organisms in the water, mapped their every gesture. Henry later wrote that although he had bathed in phosphorescent water before, he had not, he says, met with anything as beautiful as this:

> As we dived and swam, it lit us up with a green glow from head to foot so that dived we ever so deep, every movement of our bodies could be followed from the deck, although the surrounding water was black as ink, there being no moon and very little starlight. Having taken turn-about in watching one another perform evolutions in this green fire, we dressed – still wondering.[3]

Henry expresses a Victorian sense of curiosity about the natural world. During his recent training as an electrical engineer, he had studied the properties of electricity. Electric kettles, irons, water heaters and cookers began appearing in homes in the 1890s and with the coming of power stations, like one built at Deptford in south London in the late 1880s, the large-scale distribution of electricity was beginning. Henry would soon be running a successful company and making a name for himself as a leading figure in the Electrical Contractors' Association.[4]

Walter, too, would come to have more than a passing interest in scientific phenomena, like ocean phosphorescence. By 1895 scientists were already researching the properties of electromagnetic radiation, and the discovery of X-rays (later the same year) eventually had an enormous impact on medical practice. And though much medical knowledge was still uncertain and many available treatments ineffective when Walter began his studies, there was a new optimism about scientific medicine. The laboratory study of bacteria had by now led to the discovery of an effective antitoxin for diphtheria, the disease that had so devastated the CMS children's home in Limpsfield just a few years before. The public were becoming less sceptical about doctors' professional claims and therapeutic practices. Quacks, bonesetters, herbalists and other unqualified practitioners still competed with them, however, and it was not until the early twentieth century that the profession was able to restrict medical practice to members of its own ranks.

Guy's Medical School was expanding when Walter arrived. The medical curriculum had recently been extended to five years, and now the school needed more teaching and residential accommodation. New laboratories were also required in the wake of scientific advances in chemistry, physics and bacteriology. Fortunately, from the late 1890s to the outbreak of the First World War, there was a steady flow of gifts and endowments to the school and building proceeded rapidly. During Walter's time, a library and a museum were built and both the Anatomical and the Biological departments were extended. Guy's has continued to expand over the centuries, but many of its original buildings still stand. In the photograph in Illustration 10.1, taken sometime after Walter had left the school, the man in a laboratory coat on the right probably has a bottle of a chemical reagent on his shoulder which he is taking round to laboratories at the back of the building. And the man in the central arch may well have a patient in a wheelchair.[5] London is less foggy now, of course, but Guy's famous colonnade remains, and trees still line the park beyond.

Throughout the nineteenth and twentieth centuries London remained the most expensive place in Europe to study medicine, and students and their families were expected to shoulder the full cost of training. Walter was fortunate in being able to live at home, in a three-storey Victorian villa, in St Peter's Road, Croydon, not far from Coulsdon in Surrey. Between 1887 and 1894, his mother Emily lived there too, having come to England to seek treatment for her arthritis. By the time Walter entered medical school, however, she had returned to Canada to be with William in the mission field. Walter and several of his siblings continued to live in the Croydon house but as the medical curriculum expanded, the schools put up their fees. Even without residential

Illustration 10.1   A view of Guy's Memorial Park, taken from the colonnade in 1925

costs, many students were forced to borrow from family and friends and worried constantly about how to make ends meet. In all likelihood, family friends subsidised Walter's studies, just as they funded his older brother, Herbert, who went to Cambridge in the same year.

The cost of medical training was manageable only for a frugal and industrious young man. In the 1890s, dressed in their three-piece suits, and wearing rounded collars that were stiff and high, students worked hard. In some London medical schools the day began as early as 7 a.m. and did not finish before 7 p.m. You might also have to attend classes on a Saturday. But institutions had to be flexible too, to prevent their fee-paying students from leaving, and Guy's staff tolerated some lampooning. A sketch published in the hospital *Gazette* in 1900, the year after Walter arrived, has a fictional 'Demon Dean' explaining to the father of a prospective student that they will gladly teach his son – provided he knows all the subjects already:

> My dear sir, are you not aware that for your son to be admitted to Guy's he must already be a Fellow of the Royal College of Surgeons and a Doctor of Medicine, and that before being allowed to join any of our classes he will be required to write an original treatise on the subject in which he is about to be instructed?[6]

Walter took many courses with titles that today's medical students would recognise, even if the contents have altered – including anatomy, physiology and
pathology as well as chemistry, biology and physics. Courses such as 'insanity'
and 'diseases of women' now have more specialised (and less unsettling) titles
within the medical curriculum but even in the late nineteenth and early
twentieth centuries, scientific and medical knowledge was growing apace.
Students had to attend more lectures and demonstrations, study more subjects
and take more examinations than their predecessors.

While Walter was studying at Guy's, Grace was also in London, visiting factories, writing her reports and attending meetings. Perhaps they snatched time
together here and there, covertly and always with an eye out for any observers,
but we have no way of knowing. When Walter's son, Colin, came to England
many years later, he learned that one of his father's colleagues at Guy's had
known of the affair,[7] but he could not trace the man who seems in any case to
have kept his own counsel. If Walter's relationship with a married woman had
become common knowledge, it would certainly have harmed his prospects.
The profession was mindful of its still marginal status, and although a few
decades earlier London's medical students had a reputation for riotous and
irresponsible behaviour, now expectations had shifted and they had become
more disciplined and better integrated.

With the growing emphasis on uses of science in medicine, universities were
keen to have more involvement in the training of doctors. Academic medicine
was on the rise. Not all doctors were happy about the changes, however, and
tensions persisted at Guy's throughout the period of Walter's studies between
clinicians and academics, between those who thought that it was better for
doctors to learn at the bedside and those advocating lectures and scientific
learning. Despite the disagreements, the observation of skilled and experienced
colleagues at work in the ward, clinic and operating theatre was then (as now)
a vital component of training. Like some of his fellow students, Walter held
various training posts at Guy's as a 'clerk' and a 'dresser'.[8] Students were carefully selected for these positions which involved working alongside established
physicians and surgeons in hospitals as their interns or apprentices – keeping
records, passing instruments, and bandaging wounds. Once you had been
picked out of the crowd in this way, your career prospects were good.

When Walter was studying at Guy's, surgery was just beginning to emerge as
a profession in its own right. Previously, surgeons had to seek approval from
physicians before operating; now they were beginning to do things no one had
imagined possible. Anaesthesia had been introduced in the 1840s and they
could now work more slowly and complete more operations. Patients had no
longer to suffer the agony of limb removal whilst fully conscious and dressers no longer had to hold them down on the operating table – but hospitals

were still dangerous places. There were fewer deaths following surgery after antiseptic practices were introduced, thanks largely to the efforts of Joseph Lister in the 1860s and 1870s,[9] but deaths did still occur and it was several decades before consistent procedures for preventing infection were universally adopted. In the photograph taken in the operating theatre at Guy's in 1890, shown in Illustration 10.2, the doctor at the patient's head is administering an anaesthetic (probably ether). The surgeon stands to the right, ready at his side, sleeves rolled up. His assistants, his dressers and theatre nurses wear protective aprons over their outdoor clothes, but there are no gloves to be seen and no one is wearing a mask. The two metal buckets in the foreground contain carbolic acid (which Lister had introduced as an antiseptic) and despite their composure, the staff here will be suffering from its effects. Victorians became used to holding themselves steady for the camera's long exposure, but eyes will be smarting and hands stinging as they wait, draped in a toxic mantle of invisible fumes.

Illustration 10.2   The operating theatre at Guy's Hospital, 1890

During his training Walter learned about other hygiene strategies, too. William Arbuthnot Lane, a skilful and controversial surgeon at Guy's (and one of Walter's teachers) pioneered a 'no-touch' technique to help prevent post-operative infections. When he eventually became Lane's house surgeon in 1906,[10] Walter followed this procedure. The rule was that none of the surgeon's fingers or any of his assistants' fingers should go within four inches of the wound. Nothing that went into the wound either, such as a swab or ligature, was to be nearer to your hands than four inches. Lane, dextrous and meticulous, developed special, long instruments to enable surgeons to work without touching the tissues. When surgical gloves came into use he treated them simply as an added precaution and did not abandon his own methods.

Walter sat with other students (often dozens of them) in the tiers of wooden benches in a small amphitheatre, like the one in Illustration 10.2. From their high seats, rising in a horseshoe from floor to ceiling, in a circle of daylight falling from a glass dome in the roof, the students could look down on proceedings – but not always comfortably. In 1897, 'a first year's man' complains in a letter to *The Guyoscope* that he can see little. He has tried to observe several operations, including one for removing intestinal worms through the oesophagus, when he was quite unable to see the hooks which (he was told) were let down into the child's gullet. He asks, 'Why could not opera glasses, on the principle of "put-a-shilling-in-the-slot" be fitted to the back rows of rails?'[11]

Not all first year students coped with such equanimity. In a short story called 'His First Operation', published in 1894, the writer Arthur Conan Doyle (an admirer of Lane's and himself medically trained) suggests that, even after the introduction of anaesthesia, a student's first visit to the operating theatre might be a taxing experience:

> The novice, with eyes which were dilating with horror, saw the surgeon pick up the long, gleaming knife, dip it into a tin basin and balance it in his fingers as an artist might a brush. Then he saw him pinch up the skin above the tumour with his left hand. At the sight his nerves, which had already been tried once or twice that day, gave way utterly. His head swam round and he felt that in another instant he might faint. He dared not look at the patient. He dug his thumbs into his ears lest some scream should come to haunt him, and he fixed his eyes rigidly upon the wooden ledge in front of him. One glance, one cry, would he knew, break down the shred of self-possession which he still retained. He tried to think of cricket, of green fields and rippling water, of his sisters at home – of anything rather than what was going on so near him.[12]

If witnessing surgery was difficult, experiencing it was far worse, of course. Morphine had been discovered but social anxiety about its dependence and abuse led to restrictions on its distribution. The First World War would soon stimulate further pain research (as soldiers returned home with complex injuries), but in preceding decades, the management of pain left much to be desired. Mr Crosby was a patient at Guy's who underwent surgery for a fractured kneecap in the early 1880s and kept a diary of his experiences. On the day of his operation, he waited several hours: 'When I was taken in', he writes, 'I was told to look at the lights, just to take my attention off the place. I noticed all kinds of things which I trust I shall never see again. The India Rubber bag[13] passed under my nose and I was asked not to breathe. Soon I found a strange sensation possessing me though one which sent me sound asleep.' Mr Crosby knew nothing then until he awoke and found himself in his own bed 'screaming and crying'.[14]

Despite the physical and psychological stresses of medical training, there was no shortage of recruits, and enrolments in London's medical schools rose steadily until the last decade of the nineteenth century. British medical students now became unhappy about the large size of their classes and about the impersonality of their relationships with staff. They wanted the kind of informality and the closer ties that existed in other countries. Walter however, seems to have negotiated his relationships with staff particularly well, partly perhaps because he was older than his peers, and by 1907 he was being described in lustrous terms. Cuthbert Golding-Bird, a senior surgeon at Guy's, wrote a testimonial for his former dresser, now friend:

> Making permanent friends wherever he goes, he possesses gifts that make him also a persona grata with everyone, and not the least with superiors and colleagues.[15]

Other colleagues were anxious to support Walter too, and they heaped praise on one of their own, a 'Guy's man' as he would have been known. A Victorian medical student learnt loyalty (to his profession, colleagues and training institution). Discretion, tact, and skill in personal relationships were more important attributes for an aspiring medical man than intellect or knowledge; to be successful you had to inspire confidence and act decisively. Four times, Walter's 'tact' was noted in his testimonials, most significantly by Lane himself. Walter was, Lane says, 'always remarkably tactful, performing his several duties in the pleasantest way possible'.[16]

Lane was already attracting criticism from London's medical world and working closely with him would have required 'tact' of the first order. His house surgeons would later speak of his sarcasm, that some found cruel, others kindly. Mostly, he was remembered for his great surgical skill and for being 'the only man in London who could open the abdomen safely'.[17] One of his students recalled:

> He was original and unorthodox in his teaching; he used to tell us not to believe anything in text-books, that authors copied statements from other books, and he tried to get us to observe and think for ourselves, taking nothing for granted. This may not have been good for exam purposes but I think it encouraged any originality that the men had.[18]

Lane liked to invent new procedures and did not like teaching or doing routine operations. Apart from developing the 'no-touch' technique, he became world famous in four distinct fields of surgery: fractures, the middle ear, the cleft palate, and the abdomen. In the late nineteenth and early twentieth centuries it was still possible for one man to innovate across many specialisms but it was, all the same, quite an achievement and Lane was amongst those pioneering many lasting developments. Some of his instruments for operating on factures, for example, are still in use. He was colourful and charismatic and he mesmerised his house surgeons so that most thought they were going out into the world to cure all its ills. Did Walter fall under his spell? Or does Lane's emphasis on his younger colleague's 'remarkable tact' suggest a gentlemanly disagreement?

There were some who disagreed with Lane, particularly over his treatment of constipation. Lane's belief was that once the colon became overloaded, its contents became toxic and caused internal poisoning. He referred to this as 'chronic intestinal stasis', and argued that it caused internal structural changes and was responsible for a range of ailments, including general debility, rheumatoid arthritis, tuberculosis and cancer. He advocated liquid paraffin as a laxative (and used this himself three times a day), but in other cases he proposed a more radical solution. Increasingly, from 1900 onwards, he removed entire colons from constipation sufferers. Unsurprisingly, patients reported a marked improvement in their symptoms.

More than a hundred years later, the deficiencies in Lane's thinking and methods may seem ludicrously obvious, but in his time there were huge epidemics, often spread by excrement and filth, and Victorians became obsessed with the bowels which they regarded as unclean, even dangerous. Medical thinking about parts of the body and their various functions was heavily underpinned by cultural and social beliefs. For example, early Victorians thought of women as smaller versions of men, with the outside turned in (that is, with internal rather than external sexual organs) and their ovaries, even perfectly healthy ones, were sometimes enthusiastically removed because of a contemporary view that anything wrong with women must be attributable to their sexual organs.

Some were strongly opposed to the practice but in the late Victorian era, when women like Grace were asserting their right to education and the vote, many men felt threatened and few questioned the idea that women's biology made them unfit for a life outside the home. Removing the ovaries could 'help' women to accept their exclusion, for example, from the professions and the workplace. Some women colluded with the prevailing orthodoxy and actively sought the surgical removal of their normal ovaries, even before the days of anaesthesia. Eager patients like these were attacked with vitriol by the playwright George Bernard Shaw in 1911, in his 'Preface on Doctors':

> There is a fashion in operations as there is in sleeves and skirts: the triumph of some surgeon who has at last found out how to make a once desperate operation fairly safe is usually followed by a rage for that operation not only among the doctors but actually among their patients. There are men and women whom the operating table seems to fascinate: half-alive people who through vanity or hypochondria, or a craving to be the constant objects of anxious attention or what not, lose such feeble sense as they ever had of the value of their own limbs and organs.[19]

By this time, Shaw was an established member of the Fabian Society in London and a close associate of the Oakeshotts, Joseph in particular. He was ahead of the medical profession in asking ethical questions that many doctors preferred to ignore. His lengthy Preface formed part of a wider Fabian assault on government plans to introduce compulsory national health insurance,[20] and the play it was designed to introduce ('A Doctor's Dilemma', first performed a few years earlier in 1906), was a fierce satire on contemporary medical practice.

In it, one Cutler Walpole is accused of performing costly but useless operations solely for personal gain. Shaw based his character not on Lane but on a prominent throat specialist of the time who had developed an expensive and pointless procedure to remove the uvula. Lane himself was never accused of acting for ulterior motives but he did have contemporary critics, including some who were his closest associates. William Hale-White had known Lane from their student days at Guy's and was his friend; by the time Walter arrived to begin his studies, Hale-White had been appointed as a hospital physician. Like Lane, in 1907, he wrote a testimonial for Walter and like Lane, he noted Walter's personal qualities, his 'exceptional judgement and tact'.[21]

Hale-White greatly admired Lane's surgical skill, but increasingly he opposed the operative treatment of constipation that his friend was pioneering. He was not alone in doing so. Disagreements finally surfaced in 1913 at the Royal Society of Medicine, at a series of meetings which became known as 'The Great Debate'. Hale-White gave a carefully planned opening address that was polite – and devastating. In all, about fifty practitioners spoke, some for and some against Lane. Lane himself talked at length about his patients' gratitude but offered little other 'evidence' for the effectiveness of his treatment. The outcome was a clear rejection of major surgery for constipation. The meeting reached its conclusion without hostility, and arguments were countered with the greatest tact. Hale-White and Lane drove away from the meeting together. Lane, apparently crushed, simply remarked to his friend, 'I never could convince you.'[22]

It was 1905 (the same year that Grace was appointed to the LCC) when Walter completed his medical training. He passed the conjoint examination that gave him entry to both the Royal College of Surgeons and the Royal College of Physicians, represented by the letters MRCS and LRCP. He could have taken a hospital post or entered private practice. Many of his peers did both. Nineteenth-century medical schools often employed their own students and the warmth of the testimonials from Walter's colleagues at Guy's suggests he would have fitted in well. He is described by the physician E. Cooper Perry as 'loyal to authority' and 'well-liked by those with whom he worked'. 'He was excellent', says Lane, 'both as a very skilled operator and as a thoroughly scientific surgeon.' Another eminent surgeon, Alfred Fripp, describes Walter as one of the best house surgeons he has ever had.[23]

But although the young Walter had undoubted promise, men of limited means faced a struggle to establish their careers, despite their training and qualifications. Junior hospital posts involved long hours of work and poor pay; promotion was slow. Most hospital doctors needed to supplement their incomes with private work, but to set himself up independently any young practitioner required funds to live on in the early years, to pay rent and buy the drugs he would need to stock. You also needed money to purchase the accoutrements of respectability (like a carriage and the proper clothes); without these, you could not hope to attract fee-paying patients. And since women were reluctant to seek medical advice from a bachelor, a wife was also necessary.

In March 1906, while he was working as an assistant house surgeon at Guy's, alongside Golding-Bird and Fripp, Walter received news of his mother's unexpected death at Athabasca Landing, in the remote Northwest Territories of Canada. Emily had suffered from severe arthritis for some time, and had sought medical treatment in England more than once, but her death must have come as a shock. Did Walter blame his father for the accident with the buggy? Did his mother's untimely death perhaps prompt him to make changes in his own life? Whatever the truth, in early 1907 Walter told his senior colleagues that he was applying for the colonial medical service, and wished to work abroad, possibly in the Straits Settlements (now Malaysia). There was no unified colonial service at the time but professional networks were being created in India, South Africa, America, Australia and New Zealand and some of these centred on the London medical schools. *The Lancet* regularly carried advertisements for medical posts in the colonies and those recruited shared a belief in service to others. Often, too, they had a sense of adventure and, like Walter, came from religious backgrounds.

Walter made a good impression in his interview at the Colonial Office.[24] His colleagues at Guy's sent their testimonials and a boyhood friend from Croydon, Stanley Carr, who had accompanied Walter on a sailing trip with the Cash family in 1900, also wrote warmly in support. Carr's sister, Anna, had attended Croydon High School for Girls, and was in the same year there as Grace. Later, Anna would marry Walter's older brother, Herbert. The Reeves and the Carrs had lived in neighbouring streets in Croydon for some years. Now, Stanley Carr praised his friend's self-reliance and sound common sense. Walter was, he said, scrupulously honest and strictly temperate. His application having been accepted, in June 1907 Walter was offered a post with the Colonial Medical Service in Fiji – only to have the offer withdrawn a few days later on medical advice. No details are given in the official papers.[25]

It was now only two months before Grace would stage her own death, leaving her clothes on the beach in Brittany. The unhappy Victorian marriage was not easy to escape but did Walter and Grace weigh up the options together? Did they talk about the best way to preserve respectability? Once Walter had been rejected for the post in Fiji, events moved quickly. Probably through his professional contacts at Guy's, he heard about a medical locum position in Wellington, and he applied. New Zealand was a very different proposition from Fiji. It had a less taxing climate; European settlement was far more advanced but the country was just as remote. The journey would take some months and there was no time to lose.

Part III

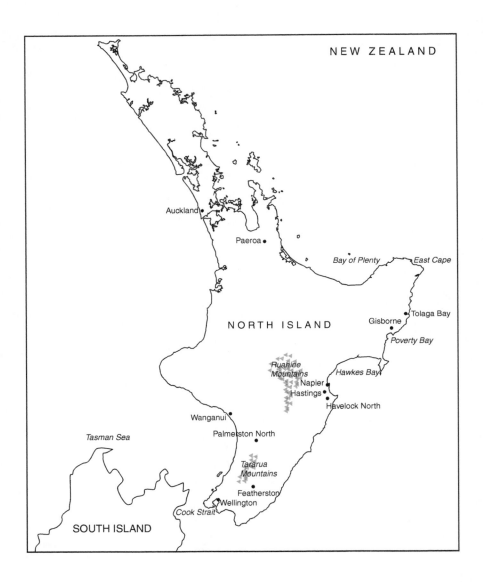

11
A Place to Begin Again

After Grace has left her clothes on a beach in Brittany, she and Walter embark on a new life together. They journey to Australia and then on to New Zealand which, at the time, is being actively promoted as a 'paradise'. Walter works briefly as a medical locum in Wellington and Grace makes her first formal appearance as his wife.

In the summer of 1907, after leaving her clothes on the beach in Arzon, Grace is believed to have swum out into the bay. She did not have to go far before she was out of sight around the headland. She was picked up by (unknown) friends and reunited with Walter a short time afterwards. The two of them then made their way to Marseille, probably by train. From here, one month later, they sailed to Sydney (via Freemantle) in Australia and then on to New Zealand.

People who are intent on disappearing take care, of course, to leave as few traces as possible. The first time Grace deliberately falsified the official records was when she boarded the *Orotava* in Marseille and gave her name as 'Mrs Reeve'. It was a pivotal moment. From that point on, whenever personal details were asked for, the couple took the opportunity to create confusion. When they were leaving France, in a close approximation to the truth, 'Dr and Mrs Reeve', gave their ages as 31 and 34 respectively. But when they boarded the *Manuka* in Sydney, they reversed the age difference. Now, Walter gave his as 28 and his companion's age was recorded as 25. The misinformation logged on the *Manuka*'s passenger list that day placed Walter and Grace in what their daughter would later call a 'happy married relation' and it gave Walter the kind of age advantage that husbands typically enjoyed at the time. Perhaps they felt the age adjustments would help them 'pass' as a conventional couple; perhaps they didn't think about it much at all, and entered the fictional details in a moment of whimsy. Or was it that they were concerned to muddy the waters for any who might try to follow?

By the time the couple landed in New Zealand more details had been invented. Grace had changed her first name to 'Joan' and adopted a fictitious second name ('Leslie'). She seems to have chosen the names at random; they have no obvious connection with people in her earlier life, and no 'Joans' or 'Leslies' have emerged from the historical records. The deliberate obfuscation of her real identity continued up to and even after her death. And there are still gaps in her story, beyond the bare facts of the case. We do not know how she was feeling in her last days at Downside Cottage, for example, or what thoughts she might have had for those she was leaving behind.

Two years later the controversial writer H. G. Wells, who was known to the Oakeshotts through the Fabian Society, published a novel called *Ann Veronica*. The work was inspired by events in Wells' own life, principally his affair with the young Amber Reeves on whom he bases his heroine. In defiance of social convention, the fictional Ann Veronica elopes with her lover to escape a marriage of convenience. The day before her planned departure, hardly daring to think of the catastrophe she is about to cause, she contemplates the things she will miss:

> She got up early, and walked about the garden in the dewy June sunshine and revived her childhood. She was saying good-bye to childhood and home and her making; she was going out into the great multitudinous world; this time there would be no returning. She was at the end of girlhood and on the eve of a woman's crowning experience. She visited the corner that had been her own little garden – her forget-me-nots and candytuft had long since been elbowed into insignificance by weeds; she visited the raspberry-canes that had sheltered her first love affair with the little boy in velvet, and the greenhouse where she had been wont to read her secret letters.[1]

At the age of 35, Grace was significantly older than Ann Veronica and, unlike her, she left behind a settled adult life and a career. In 1907, Grace's parents (now both getting on in years), as well as her two sisters and brother, lived only a few doors away from her in Fanfare Road. She must have known it would be impossible for her ever to see any of them again. The family of George Drysdale (who had staged his own death some decades earlier) were more fortunate. The brilliant young Scotsman (who became a free-thinker and advocate of contraception) left his clothes on the banks of the River Danube in 1844, when he was on a walking tour of Europe. His family feared the worst and announced his drowning in the newspaper. Then, about two years later, he reappeared. Eventually, it turned out that he had wished to escape the shame of a 'sexual neurosis' (an adolescent obsession with masturbation) and to seek help for it in secret. Publicly, his family were overjoyed at his return, though, as the writer Kate Summerscale observes, they must have experienced some hurt and confusion, too.[2] In another case, the family of Gabriela Cunninghame Graham seems to have kept in touch with her despite her adoption of a fictitious identity in the 1870s. Again, the motive for the deceit was the avoidance of shame, probably a past life as a prostitute.[3]

Like Graham, Grace would quickly have become used to living with her secret. Before long, it may not have felt like a secret at all. She had found a way to avoid the shame of becoming known as a 'fallen woman' and the huge expense of divorce. Her professional reputation was intact; respectability had been preserved, at least so far.

The Tasman crossing from Australia to New Zealand could be rough. When Walter and Grace joined the *Manuka* in Sydney, the steamer was carrying nearly two hundred passengers and had accommodation for twice as many. Built in 1903, the *Manuka* weighed four and a half thousand tons and had a speed of 15 knots, but even on board the larger steamers passengers might find themselves in distress. There are over a thousand nautical miles between Sydney and Wellington and although this particular sailing was less crowded than usual, conditions on board were often cramped. A young child who made the voyage from Australia in 1906, on a sister-ship called the *Maheno*, in later life recalled 'people lying on the floor or any available space, most of them terribly sea sick'.[4]

The Great Migration of the 1870s was over by the time Walter and Grace made the crossing, but interest in the far colony of New Zealand had revived following the introduction of refrigeration. Dairy markets had opened up overseas. Jobs were available and the colonial government was once again offering assistance to migrants. But not all of Grace's and Walter's fellow passengers were from Britain. Between 1900 and 1915, about a third of New Zealand's migrants came from Australia and in the five years from 1901 to 1906, the number of Australian-born residents in New Zealand almost doubled.[5] Drought and economic depression encouraged them to leave the continent – and fares across the Tasman were low.

Those travelling in saloon cabins paid higher fares and enjoyed more space, privacy and better food than those below in steerage. Walter and Grace travelled in saloon class with other middle-class passengers, including several families. Steerage, a low-ceilinged space beneath the main deck, was for the labouring classes, and most of those migrants were seeking work on farms or in domestic service. The names of those in steerage did not appear in newspaper reports but a 35-year-old Australian socialist and trade unionist, Michael Joseph Savage, was also on board the *Manuka* that day, having purchased a ticket for only 75 shillings.

Saloon-class passengers did not usually mix with those from below decks but if Walter and Grace had spoken to him, they might have found they held similar views about the way a fair society should be organised. Savage and some of his compatriots brought new political philosophies to the young country and a radical, American-influenced approach to industrial unionism that would lead to the formation of the Red Federation of Labour (commonly known as the 'Red Fed'). Savage would eventually become the leader of New Zealand's first Labour government and the most loved of all the country's Prime Ministers.

New Zealand had by now become a magnet for many seeking a better way of life. During the nineteenth century it had been promoted as a kind of paradise, and even then the country was known for its exotic environment. It was a place where any intending immigrant might expect to better themselves. On a visit in 1872, Anthony Trollope, the novelist, was struck by the lack of social deference:

> The very tone in which a maid servant speaks to you in New Zealand, her quiet little joke, her familiar smile, her easy manner, tell you at once that the badge of servitude is not heavy upon her. She takes her wages, makes your bed and hands your plate – but she does not consider herself to be of an order of things different from your order. Many who have been accustomed to be served all their life may not like this. If so, they had better not live in New Zealand.[6]

Unlike the penal colony of Australia, New Zealand was still thought to have utopian potential and its remoteness from Europe, its beautiful landscapes and healthy climate all fostered this perception. In 1900, about a thousand members of the working class Clarion Fellowship, from Manchester in the north of England, had travelled to the young colony expressly to create a socialist community there.[7] The country had a reputation for radical political experiments and, supposedly, a better history of contact between its European settlers and its indigenous people. Newcomers hoped to build a society that avoided the pitfalls and mistakes that had plagued the Old World, particularly the social class tensions and the extremes of wealth and poverty. Michael Joseph Savage had seen these portrayals of New Zealand in Australian newspapers and journals, and like many others he had made up his mind to see it for himself.

Grace and Walter would also have had known a little about this small, isolated colony in the southern ocean. In the early 1900s Walter had had the chance to meet young New Zealanders at Guy's Hospital where they were regularly recruited onto courses in medicine and dentistry. He may well have used these contacts to secure his locum post in Wellington, as well as to find out more about living conditions. And Grace would have known, for example, that once in New Zealand she would be able to vote, since women there had won the franchise in 1893.[8] She may also have known about the introduction of old age pensions in 1898 and, through her work for the Women's Industrial Council, she would certainly have known about the New Zealand Industrial Conciliation and Arbitration Act of 1894. This legislation protected unions and required the peaceful arbitration of industrial disputes.

In addition, some of Grace's professional associates in London were women with favourable first-hand views about life in the new colony. Beatrice Webb had visited New Zealand in 1898. She commented on the dowdy appearance of local people with 'their mackintoshes, umbrellas and big thick boots'.[9] The political system was, she said, 'crudely democratic and unscientifically collectivist',[10] but she loved the bush 'with its towering Kauri trees and its luxuriant undergrowth of tree ferns, creepers, mosses, all tangled up together in a variety of yellow, brown and green strands'.[11] At the end of her visit, she concluded that 'taken all in all, if I had to bring up a family outside of Great Britain I would choose New Zealand as its home'.[12]

Another of Grace's colleagues was Maud Pember Reeves, wife of the New Zealand statesman and writer (and the mother of Amber Reeves). In 1899 Reeves published an article in the WIC's newsletter about working women in New Zealand. In it, she explained that because the men there were better paid, fewer women worked for wages than in England. Further, women were in the minority and most were required in the home. Where women were employed outside the home, in factories for example, their hours were limited and they had better working conditions.[13] Webb and Reeves were friends and they were connected to Grace in another way too. Both were leading members of the Fabian Society in London, an organisation in which Harold's older brother, Joseph, was particularly prominent. By marrying into a Fabian family, although her name does not appear in membership records, Grace had had many opportunities at meetings and public lectures to rub shoulders with other Fabians, and Oakeshott family gatherings would also have attracted the Society's notables. Moving in these circles, she could not have escaped hearing about social reforms in New Zealand.

During the early years of the twentieth century, an army of radicals visited the colony to test the paradise myth for themselves, including Ramsay MacDonald, who came in 1906 with his wife, Margaret. For the most part, these visitors had returned full of enthusiasm for the country's potential.[14] From Grace's perspective, however, New Zealand had one other important thing going for it, and that was its sheer isolation; it was, quite simply, as far away as you could go from England and for the construction of a new and anonymous life, it was the ideal choice.

The voyage across the Tasman from Australia took four days. On 9 October the *Manuka* anchored in the stream at a quarter to one in the afternoon and it berthed at the quay in Wellington about an hour and a half later. In the glassy, southern light, the hills behind the harbour stood like cardboard cut-outs. Clusters of wooden buildings clung to windy slopes. 'Mr and Mrs Reeve' disembarked. It was spring. After a pleasantly warm and fine morning, the weather had turned and soft drizzle greeted the travellers.

Illustration 11.1   Overlooking Wellington City, 1905

Wellington was a prosperous port. Domestic servants were in demand and local middle-class matrons keenly awaited the arrival of the immigrant ships. Some even boarded them at the quayside to recruit single women as house-keepers, nurses and maids. Men were needed as farm labourers. Commerce was growing rapidly and shops and banks were springing up along the city streets. Just weeks earlier, the country had become a Dominion but 'The Empire City' (as it was called) had strong links with Britain and aspirations to match. In 1901 the combined population of the city and its suburbs was nearly 50,000 and it would double by 1921.[15] The mood was optimistic. New Zealand's first cars had come to Wellington in 1898 and would soon be commonplace. Electric trams had arrived in 1904. The horse-drawn tram service was abandoned a few weeks later but horses remained on the city streets until the 1920s. Early on Sundays, stable owners took them to nearby Oriental Bay for a swim in the harbour – and local residents seeking a dip in the waves were obliged to wait their turn.

As it happened, the day Walter and Joan arrived was Labour Day and a pub-lic holiday. Offices and banks were closed and over 8,000 workers and their friends were gathered at Days Bay for celebrations. This sandy beach is on the eastern side of the harbour and sports, music and speeches were in progress, despite the advancing rain. The newly-arrived Savage had only one shilling in his pocket when the *Manuka* docked, and he spent half of this on a shave. After seeing a poster he made his way to Days Bay and there listened to Ben Tillett, a visiting British radical, chiding New Zealand workers for having less sense of their rights and power than ten years before. Savage would say years later that Tillett's voice was 'the first he heard in New Zealand'.[16]

Once the public holiday was over, Walter wasted no time in applying for registration as a doctor. Three days after his arrival, he took evidence of his London qualifications to the office of the Registrar General and placed a notice in *The New Zealand Gazette*. On 13 November 1907 his name was added to the Medical Register for the Dominion of New Zealand. A short time later, he was acting as a locum for a Dr Elliott (and his colleague, Dr Young). Both these men held posts at the local hospital in Wellington. James Elliott had been appointed as the first House Surgeon there in 1903. Like Walter, he had a clergyman father and he had been brought to New Zealand when he was four years old. And like Walter, although it was still several years away, he would eventually be caught up in the First World War, when he served as medical commander on the *Maheno*, which by then had been fitted out as a hospital ship.

In the meantime, Walter and Joan needed to adjust to their new life. Wellington could hardly have been more different from London. Most of the buildings were wooden, colonial-style structures with one or two storeys. The city sits on a fault line and the earthquake risk had encouraged early settlers to adopt safe styles of construction. But the wooden buildings in turn posed a fire hazard, and when insurers began offering lower fire premiums for concrete constructions, taller and stronger buildings began to take shape. Walter and Joan were unused to the hazards posed by fires and earthquakes, but during their time in Wellington there was a huge conflagration that must have made a lasting impression.

In the early hours of a December morning, the nightwatchman at Parliament Buildings, a Mr Amos Wilby, made his rounds before settling down with a cup of cocoa. Then he heard a peculiar noise: 'It sounded like rain on the roof', he said. 'But I could not believe that rain could be falling seeing that a little while before, the night appeared so fine.' The cause of the noise was soon apparent. The building was on fire and 'to anyone who knew how old and tindery that portion of the Parliament Buildings was the possibilities were enormous'.[17] Mr Wilby was later described as 'a sober and attentive man' who was neither a smoker nor a drinker,[18] and Ministers were satisfied that he had carried out his duties on the night in question. Nevertheless, the fire quickly gained hold and within the hour the old wooden wing had completely gone. That evening, local journalists dug deep to describe the scene they had witnessed. The fire 'devoured the wood as an elephant might crush a peanut', it 'breakfasted ravenously', and it billowed 'lurid smoke', 'like volumes of blood-red steam from an inferno'. It threw sparks and 'glowing fragments soared upwards, and then zigzagged to the ground – a shower of rosy snow'.[19]

Illustration 11.2   The fire at Parliament Buildings, Wellington, in 1907

The only section to survive was the brick Assembly Library (on the right in the photograph in Illustration 11.2). Even here, there were some anxious moments when the precious store of books and documents seemed to be in danger and dozens of pyjama-clad citizens began ferrying them to safety. In the nearby cemetery, we are told, even the dead took note. The spirit of Richard Seddon, New Zealand's longest-serving Prime Minister who had died just eighteen months earlier, awakened 'at the sight of the scene of all his triumphs during a score of years consumed to ashes in a single hour'.[20]

Later on the day of the fire, a group of women gathered for a tea party. Mrs Elliott, the wife of Dr Elliott, was welcoming Mrs Reeve, wife of Dr Reeve, the locum from England. According to 'Priscilla', the writer of the Ladies Column in the local newspaper, the tea party for Mrs Reeve was a gossipy affair, but gossip 'in the pleasantest sense of the word with no sting or malice'. 'Exquisite flowers' adorned 'pretty rooms' and the new Elliott baby was much admired. The hostess wore a charming white muslin frock and her guests enjoyed strawberries and cream in the summer sunshine. True, the 'ghastly ruins' of nearby Parliament Buildings cast a shadow and made Priscilla wonder if it might seem 'trivial' to write of 'fripperies and frivols'.[21] She did not know, of course, that the guest of honour at the tea party (presumably by now equipped with a convincing back-story) was masquerading under a false name, or that this social occasion marked the beginning a new life for Joan in circumstances that were far from trivial.

About five months afterwards, Walter and Joan left Wellington behind and boarded the steamer *Wimmera*. They sailed north to Poverty Bay and in the early hours of a Saturday morning in May 1908 arrived in the port of Gisborne. It was overcast, with a strong westerly wind. Within a few days Walter had taken over the medical practice of a Dr Cole in the town, and Joan had advertised for domestic help. She was already pregnant.

12
'Ignoble Motives'

In Gisborne, Poverty Bay, Joan and Walter have to adapt quickly to colonial life. Joan is safely delivered of twin sons and before long has become a leading figure in the local Plunket Society, set up to promote infant welfare. Walter's views about the best way to treat the victims of a poliomyelitis epidemic spark a row and soon an enquiry is set up to investigate the running of the public hospital.

> Luncheon was over, and Miss Peterson (familiarly known as 'Minnie') was engaged in the washhouse at the rear of the hotel when accused is said to have entered the shed and spoken to her. After a brief conversation, Symons is alleged to have fired at the girl twice (some say thrice), but so far as the medical gentlemen could ascertain only two bullets entered her body, one lodging at the base of her left lung whilst the other penetrated her lower jaw, breaking it and finally lodging in the base of her skull.[1]

So reads a report in a local newspaper in Gisborne in June 1908, about a month after Joan and Walter had arrived. One of the 'medical gentlemen' who attended the scene was Walter himself and the shooting was a startling introduction to a new life. The Pakarae hotel, where the incident had occurred at a little after two o'clock in the afternoon, was in a remote location on the coast, about 18 miles north of Gisborne.

A Dr Collins was first called from the town, but on the way he was forced to abandon his motor car and complete his journey in a horse-drawn vehicle. When Walter was summoned, he, too, was delayed by the rising tide and he was stranded at Tapuwai rocks for almost an hour.[2] He finally reached Pakarae at 3 a.m., and by 4 a.m. the two doctors had successfully extracted one of the bullets. But Minnie was not expected to survive the attack and a clergyman was called. Walter returned to Gisborne at 11 a.m. the following morning. Eight days later, the second bullet, the one that had lodged in Minnie's skull, was finally removed and she began to improve.

146

As Walter was discovering, Gisborne was a frontier town.[3] The East Coast had a deserved reputation for drunkenness but unlike the outposts of America and Australia, here several key social institutions were already in place. When Walter and his colleague went to the scene of the shooting, they were accompanied by a detective from the local constabulary who arrested the perpetrator and brought him back to town where he was taken to court and formally charged. Illustration 12.1 captures a similar incident. In this photograph, taken at about the same time, one Detective John McLeod from Gisborne (standing on the right-hand side of the coach) is escorting a prisoner he has arrested further up the coast. From the early days of European settlement, law courts provided opportunities for the resolution of tensions, largely containing the disorder and violence that was typical of other frontier communities.

Yet colonial life could still be physically challenging, especially in isolated places, and Walter and Joan had to adapt quickly. By now winter was setting in and snow had fallen in some districts. And it was not just the seasons that were topsy-turvy. London, Croydon, the soft slopes of Ranmore – all were far away and though the European migrants had brought their institutions with them (the hospitals, courts and schools were familiar enough) there were differences, too. Thanks mainly to a rapid growth in wool exports, and hopes that oil would soon be discovered, Gisborne was a boom town. The population rose from 2,737 in 1901 to more than 15,000 in 1926.[4] But the town's infrastructure was 'unfinished'[5] and, as Walter discovered on the night of the shooting, travel

Illustration 12.1 Redstone's coach travelling north of Gisborne in about 1910, possibly on Makorori beach

could be hazardous. A few months after the Pakarae incident, he was lucky to escape unhurt when his horse bolted in the centre of town and his gig collided with a tree. The following year, he imported a Grégoire car from France (named after its manufacturer) and clearly enjoyed driving it. More than once he was fined for speeding but he was still obliged to make some journeys by horse and cart.

Few local people owned motor cars. Gisborne had no sealed streets and no footpaths and the main Gladstone Road was a muddy bog in winter, a dusty desert in summer. Elsewhere in the region, the beaches and river beds at times served as roads and for longer trips the sea was still a highway. Small coastal steamers, like the one that had brought the couple from Wellington, were vital for transporting both goods and people. The local terrain consistently defied attempts to complete a railway and it was not until the early 1940s that you were first able to take a train from Gisborne to Napier and Wellington.

It was not just travel that was difficult. Many ordinary activities were more arduous in Gisborne than had become usual elsewhere in New Zealand. When Walter and Joan first arrived they had no electricity, and gas was used to light the streets and houses. Although telegraph and telephone facilities existed there was no water supply or sewerage system. Lack of amenities in the town led to outbreaks of enteric diseases and the occasional epidemic. In 1900, in an effort to counter the spread of bubonic plague, a bounty of 3*d*. was offered for every rat that the locals managed to kill.[6]

Most migrants would have hoped they were leaving diseases like bubonic plague behind them in Europe. And usually they did experience health benefits in New Zealand as their diets and living conditions improved.[7] As Beatrice Webb had observed after her visit to the country in 1898, all things considered, it was a good place to bring up a family.[8]

In August 1908, about three months after her arrival, Joan gave birth to non-identical twin sons, Antony Walter and Colin David. During childbirth she would have been at risk, particularly from infection. Until the early 1920s, Pakeha women (that is, non-Maori New Zealand women who are mostly but not exclusively of European descent) who were of child-bearing age had a higher death rate than men of the same age.[9] And new migrants like Joan, were typically without their extended families, their parents and grandparents. It was as if the generations were truncated in the process of colonisation, write Porter and MacDonald in their collection of women's writings from nineteenth-century New Zealand. For most settlers there was no network of relatives to fall back on.[10] For Joan, of course, the very point of going to New Zealand was to break her existing ties. And she had hardly been in Gisborne long enough to build new friendships. She was without her mother and sisters, and she was expecting twins, but fortunately Walter was a doctor.

Unlike his own father who had helped bring him into the world more than thirty years before, Walter had successfully completed courses in midwifery when he was studying at Guy's and now, in this remote frontier town, his professional training was invaluable.

During childbirth in early New Zealand, women who were without their families often turned to neighbours and friends, like this letter writer in correspondence with her mother some years before:

> My kind friend Mrs Wilson did everything for me that you could have done; she came every day to dress the baby until I was strong – in short, as I often told her, she was both mother and sister to me.[11]

Medical help was particularly scarce in the less accessible regions of the country. Two years later Walter would be called to attend a shepherd's wife in an inland settlement on the banks of the Waipaoa River, about twenty-eight miles north of Gisborne. This time, he was able to go by train but still it took all afternoon and most of the evening to reach her. By then, the 32-year-old woman was very weak. A miscarriage had resulted in a haemorrhage and despite Walter's best efforts she died in the early hours of the following morning.

The Reeve twins, however, were delivered safely and Joan's domestic life was now made easier by paid helpers. Within days of her arrival she had advertised for a good general servant, and a few months after the twins' birth she was seeking a young girl to assist with housework. Both she and Walter (and Harold, too) had grown up in families with one or two domestic servants, and there was no shortage of help in the new dominion. She would quickly have discovered, however, that social relationships were differently configured here, that people were less status conscious and interactions less formal. 'You can't bring drawing-room conventions here', says a character in Jane Mander's 1900s novel, *The Story of a New Zealand River*. She is advising a younger woman, a new colonist, in an isolated settlement in the North Island. 'And when we English people find ourselves away in places like this we can't afford to snub each other because of a difference in the work we do. We drop all that when we leave England.'[12]

The migrant ships had brought people to New Zealand from many different social backgrounds but the one thing the new arrivals had in common was their optimism and their belief that here they could better themselves. In Gisborne, when Joan and Walter arrived, the local community was youthful and masculine,[13] and it was overwhelmingly Pakeha. By 1926 fewer than two per cent of the town's population were Maori[14] and the couple would initially have had little contact with indigenous people.

Within six months of their arrival, however, the Reeves were living in a district that is still known as Kaiti (on the eastern side of the Turanganui

River, across a bridge from the main part of town) and here, in contrast to the township itself, there was a strong Maori presence. Kaiti was developing into a highly desirable suburb. The land here had been sold to Pakeha settlers by Maori in the 1880s but important local Maori had retained some blocks of land for themselves and continued to live in the area.[15]

Joan and Walter purchased a house in Wainui Road (now the main road north out of Gisborne to the Eastern Cape) on the corner with Hirini Street, and just behind it was Kaiti Hill (also known as Titirangi Reserve) where the first Maori migrants to the area had established a settlement long before. At the base of the hill, in Hirini Street, a few minutes' walk from the Reeves' place, was a Maori cemetery and Te Poho o Rawiri meeting house.[16] The family would often have passed these sites. Also nearby was the landing place of James Cook, where the first formal meeting between Maori and European took place in 1769, and a monument commemorating the event was erected here in 1906.

More than a year had now passed since Joan had left her clothes on the beach in Brittany and the couple must have been pleased to have their own home at last. It was a wooden, two storey colonial-style house with ten large rooms, as well as several outbuildings. Set in half an acre of land, it was to be their home for the next ten years. The house no longer exists, but at the time it was known as one of the best residential properties in Gisborne.

Illustration 12.2 A postcard view of Kaiti taken in about 1910 by an unknown photographer. Wainui Road runs from left to right across the centre of the photograph and the main part of the town lies away to the left, across the Turanganui River. Records suggest that the house marked here, long since demolished, was the Reeve family home

It was here that Walter held his surgery and here, in 1912, that their daughter, Renée Mary Elizabeth, was born. In naming her, Joan forged some discrete links with her past. As well as remembering her close friend at Cambridge, she now acknowledged her absent mother, Elizabeth, and her great-aunt, Mary, and so connected her daughter to a family she would never know. All the children's New Zealand birth certificates give their mother's name as 'Joan Leslie Reeve', but whereas in 1907 she had given her age as 25, in 1908 when the birth of her twins was officially recorded, she claimed to be 33. And when Renée was born in 1912, Joan gave her age as 37, although in fact she was about to turn 40. It was a game of hunt the thimble and, like the thimble, Joan was always in plain sight. Clues to her real identity were present but always randomly scrambled.

Many had less compelling reasons to migrate than Walter and Joan, and travelled to New Zealand simply to provide a better future for their children than might have been possible in their home countries. The social freedoms, better diets and healthier outdoor lifestyles, however, sometimes came at a cost. Some middle-class Pakeha parents feared that their offspring would grow up to be 'colonial'. Usually, this meant wild and boisterous, with uncouth manners, barefooted in all likelihood, independent to an undesirable extent, and with a 'pervading restlessness'.[17] You might be particularly worried about your daughters whose marriage prospects could be adversely affected. Walter, however, was determined that his children would have an uninhibited childhood, as little like his own as possible. The twins were agile and mischievous and one morning, when he was shaving upstairs in the Kaiti house, Walter caught sight of the boys, hanging from the drainpipe and peering in at him through the window.

The Reeve children would all remember their childhood with fondness. Though Joan might now have settled for a busy, domestic (and anonymous) life, instead she did what she had always done. The twins were just two years old when their mother was first elected to the committee of the newly formed Cook County Women's Guild[18] and soon, she had also joined the local library committee. Tony, the eldest of the twins, would later say that as a child he grew to hate the words 'committee' and 'meeting' – so much of his mother's time was taken up in this way. In her absence the children were cared for by hired servants and one (a nanny called Rosie who lived with the family for many years), became a particular favourite.

Joan's focus now shifted away from education as she pitched her energies into local campaigns on child and maternal health. Though maternal mortality figures were bleak, in the first half of the twentieth century New Zealanders were proud to have the lowest infant death rate in the world and determined to maintain their record.[19] In May 1907, a few months before Joan and Walter arrived, Frederic Truby King (who had given up his job as a bank clerk in

Illustration 12.3   A kindergarten class in Gisborne in about 1912. Colin is second and Tony fourth from the left

New Plymouth to train as a doctor) had addressed a meeting in Dunedin. He was a small man (later he became stooped) with a massive head and dark, arresting eyes; he had considerable charisma and a dramatic delivery. This meeting marked the beginning of what was initially called the Society for the Promotion of the Health of Women and Children (soon to become known as the Plunket Society, after the wife of the then Governor who was a keen supporter). Eventually, it would become the country's most famous voluntary organisation. Though its primary purpose was educative, the Plunket Society also provided a framework for women to come together, support each other and take part in public debate.[20] It was the perfect way for Joan to return to activism and to integrate herself into the local community.

In 1912 Dr King toured the country to promote the new society, and at the end of July he reached Gisborne. In a crowded schoolroom he lectured to a largely female audience on the importance to a growing child of fresh air, play and proper feeding.[21] Joan was amongst those who helped organise his visit and with her background on the Women's Industrial Council, and their fight for equality, she might have found his insistence on the need for a different system of schooling for girls, as opposed to boys, rather less compelling than his emphasis on what science could teach mothers. But at a public meeting a few weeks later both she and Walter lent their support to the burgeoning infant welfare movement and she took on the role of Committee Secretary.

By the end of King's tour there were 70 local branches of the Plunket Society in New Zealand, including the one in Gisborne. It was typically middle-class women who took up the cause, as an act of citizenship, and the wives of doctors, politicians and religious leaders were prominent. With the ever-increasing support of the state, women soon made the Plunket society their own. In Gisborne, as Joan took on the secretarial role, Mrs C. A. de Lautour, the wife of a prominent lawyer and businessman, was elected President and a Mrs J. Townley, the wife of a former Mayor, was made Patroness. One Mrs D. W. Coleman, the wife of a future Mayor, became a Vice-President.

Like Joan, many of the society's members were involved in a range of organisations and, as in England, such activities enabled them to participate in public life in their own right when paid employment was not an option. In London Joan had worked with a number of notable and eminent people and she seemed now to settle easily again into a productive, civic role. She was accustomed to the work of an organisational secretary; the post may have lacked the prestige of a presidency, but it put her in direct communication with local stakeholders and she used the opportunity to advantage.

In the summer of 1913, for instance, she wrote to local milk vendors on behalf of the Plunket Society. She tactfully explained that the frequent and severe epidemics of summer sickness and dysentery in the district would be greatly reduced if the milkmen could keep the milk cool while it was on its way to local families. The ever-practical Joan then specified exactly what they should do. Each milk can should be wrapped in wet sacking: 'This sacking should be kept wet during the whole round', she said. 'One milkman finds this can be managed by wetting the sacking at the drinking troughs.' Covering the cart with dry sacking encouraged the growth of harmful bacteria in the milk because the air underneath became hot and could not circulate freely. The milk vendors should also avoid mixing night and morning milk, she said, to minimise the risk of contamination.[22]

A distinctive feature of the new organisation was its nurses who were trained in infant welfare. They held local clinics and visited mothers in their homes. Their work was overseen by committee members and in Gisborne, it was Joan who advertised for a local room that the Plunket nurses might use as an office. From the outset some friction existed with the country's medical profession (keen to preserve their power and control), but many local doctors remained supportive of the women's work and Walter was an honorary medical advisor to the Gisborne society from its inception.

New Zealand's Plunket Society soon generated interest overseas, in the United States, Canada and Australia. Truby King himself was invited to England to help launch the movement there. So it was, says Linda Bryder, in her study of the politics of infant health and welfare in twentieth-century New Zealand, that 'Plunket was sanctioned from the heart of the empire'.[23]

In later decades the movement had its critics, and King's insistence on routine, the importance of breast milk and 'feeding by the clock' led to claims that he had sought to turn out self-controlled citizens, good factory workers and soldiers, who would thrive only if they learned obedience.[24] At the time, however, it enjoyed popular support. Although medically inspired it was run by women for women, and widely believed to be responsible for the reduction in infant deaths that occurred in New Zealand between about 1906 and 1936.[25]

A month after Truby King's visit to Gisborne in 1912, Joan helped stage a children's carnival to raise funds for the new Plunket society. The local newspaper described it as a 'delightful afternoon' for 'a large company of happy, smiling children'. There were 'little tots who had not long found the use of their legs', as well as older children and all were in fancy dress. Renée Reeve, aged seven months, was one of the tots – and she was dressed as the Queen of Hearts.[26] Other fundraising carnivals would follow, particularly after the outbreak of war in 1914, and Joan was always the mainstay, her energies key to a successful event.

On the evening of the children's carnival, Walter took part in a concert in the town at His Majesty's Theatre. According to the reviewer, he had 'a sweet tenor voice and clear enunciation'.[27] It was not Walter's first appearance on the local stage. In the winter of 1911 he had taken the lead in a one-act drama called *In Honour Bound*, and starred alongside Rosemary Rees (who later became a successful novelist).[28]

The play, written by English dramatist Sydney Grundy in 1880 involves a young wife, neglected by her husband, who seeks distraction and consolation elsewhere. It had been given a reading in 1884 by Eleanor Marx and Edward Aveling at a London meeting of the Social Democratic Federation, as if in a deliberate attempt, one observer remarked, to invite scrutiny of their own unconventional living arrangements.[29] The Fabian Society also used performance to encourage debate on the merits of free love, socialism and votes for women, and George Bernard Shaw wrote a number of plays in the 1890s deliberately to provoke criticism of social and romantic conventions.[30]

In early twentieth-century Gisborne, however, the motive was far more likely to be entertainment. Walter played the part of the husband, Sir George Carlyon, who toys cleverly with both wife and lover until he has the proof he needs – which he then destroys. Were Walter and Joan amused by the play? Did it prompt any feelings of guilt or recognition? Or were they by now so embedded in community life, playing their own real-life parts so convincingly that it did not even raise a smile? Whatever their private thoughts about their early days in Gisborne, they are not likely to have realised just how uncongenial life here was about to become, particularly for Walter.

As a local doctor, Walter worked within what would now be called a 'user pays' system (just as he would have done had he stayed in England). But in early Gisborne, that the user did not or could not always pay is clear from court reports of the time, and Walter is listed as a creditor in numerous local bankruptcy cases from 1908 onwards. On one occasion the owner of a hotel had been taken ill and was unable to attend to his business; other cases involved a farm labourer, a sawmill worker, and a carpenter who had been injured in a riding accident. Renée would later recall being at home sometimes and hearing a thump on the family's verandah as an impecunious but grateful patient deposited a leg of lamb in lieu of payment for medical fees.

The family themselves lived very comfortably, and as a local doctor Walter was held in high esteem by the community. In 1913 he was one of three doctors elected to the public hospital's honorary staff. Along with a Dr C. F. Scott and a Dr J. W. Williams, his role was to assist and advise the new medical superintendent. At this time public hospitals like the one in Gisborne were funded by local rates, government subsidies, patient fees and voluntary contributions. They were originally intended for poor people (and for Maori) and patients who could not afford to pay were not charged. Honorary staff (like Walter) did not receive remuneration for their contributions to the work of the hospital but they gained valuable medical experience which they could then use in the treatment of their own fee-paying patients. It was in the implementation of his honorary role that Walter came into conflict with some of his colleagues a few years after his appointment and before long he had become embroiled in hospital politics.

It began as a dispute about the best way to treat infantile paralysis (or poliomyelitis). In 1916 the country was at war – and it was also in the grip of a major polio epidemic.[31] International research had identified the virus responsible for the disease but no one knew how it was spread. Anxiety ran high and parents like Walter and Joan had to remain watchful since children and adolescents were most at risk.[32] Early indications of the disease, which included headache, fever, neck and back pain, as well as gastro-intestinal symptoms, could all be mistaken for influenza. Concrete information was scarce but some believed the onset of polio was somehow connected to exposure to strong sunlight. Those in rural areas and from more affluent backgrounds seemed to be in particular jeopardy. In February, Gisborne was the worst affected district in New Zealand with 24 cases reported. Despite her parents' vigilance, Renée Reeve, aged about four, contracted a mild form of the disease, though she suffered no lasting effects. Others were less fortunate. By July the epidemic had petered out nationally but not before it had claimed the lives of more than one hundred children.

The disease did not disappear completely; for the next 45 years there were always some cases of polio in New Zealand and at least one death every year.

A survivor of a later epidemic recalls its impact. She contracted the disease at only six months old and was put in isolation. Neighbours were in dread of her, and her family became strangers. It meant 'no more breastfeeding from my mother, no cuddles from Granny, or piggybacks from my young uncles'. When she eventually returned home, she had sisters she didn't know, a mother she had never bonded with and a father who was determined she was going to walk.[33]

Just as there was uncertainty about the spread of the disease, so there was controversy also about the best way to treat its victims. Massage was the most widely used therapy but its effectiveness was unproven. Then, on 7 April 1916, the *Poverty Bay Herald* published a leader about the treatment of polio in Gisborne's public hospital. In this small, remote community that Walter himself described a few months later as 'a backwater',[34] this local newspaper energetically took up community causes. Though the Gisborne Borough Council had begun generating power at an oil and steam station in 1912, electricity was not yet widely available and the local press was the only way most of the town's residents had of keeping up with events. During crises and controversies, those in influential positions were not afraid to use the *Herald* as a mouthpiece and they could be sure of having an impact. On the basis of information received from a certain Dr Charles Scott, a senior consulting surgeon (and Walter's colleague on the honorary staff), the newspaper now claimed there were up to twenty children in the hospital who had lost the use of their arms, legs and back. Massage was the only thing that would save them from becoming permanently disabled and this treatment was being withheld because the nurses were simply too busy.

A few days after the leader article, Walter wrote to the newspaper insisting that it had been misinformed. It was not true, he said, that the children were being neglected because nurses were too busy to provide massage treatment. The patients had not received massage because that treatment had not been ordered for them by medical staff: 'In the opinion of the majority of us in the early stages massage is not only unnecessary but actually harmful, and may do irreparable damage', said Walter. 'In this opinion we are following an eminent English specialist in this disease, who does not allow the use of electricity or massage for at least three months.'[35] Instead, said Walter, 'the cases are being treated on the principle of complete rest to the affected parts, both in the stage of complete paralysis and also in the stage of recovery of the damaged muscles, when the muscles are in a highly delicate condition and will not stand handling'. This approach was proving successful. 'On the occasion of my last visit to the isolation ward', said Walter, 'I noticed that nearly all the beds in use were empty, as the children were up and about and had recovered the use of their limbs to such an extent that it required a skilled eye to detect any defect.'[36]

But Scott was not to be so easily placated. He had the bit between his teeth – and he had a vested interest in massage treatments. His riposte must have stunned readers. He accused Walter of stirring up public unrest, of instigating a professional brawl and of launching a personal attack on him and his new 'Ladies Massage Society', whose volunteers were doing so much good. And who exactly was Reeve speaking for? Scott had telegrammed fourteen of the country's 'leading medical men' and proceeded to list their opinions about when massage should be started in polio cases. Most experts, he claimed, introduced massage as soon as possible after the end of the acute stage. To wait longer was to risk finding that 'the muscle fibres no longer exist, the process of degeneration and waste having blotted them out'.[37]

The intensity of Scott's attack must have stung; in a small community, where professional and personal relationships were bound to overlap, it was damaging. It was a far cry, too, from the gentlemanly manner in which professional disagreements had been addressed at Guy's Hospital in London during Walter's training. Now at last, the Chairman of the Hospital Board spoke up.[38] Dr Reeve was 'the last man' to take part in any 'professional brawl', said Howard Kenway. He was 'quite incapable' of publicly accusing a fellow practitioner of untruthfulness or of misrepresenting the facts.[39]

Two days later, Walter himself carefully refuted Scott's allegations: 'I would ask Dr Scott to read my original letter again. He will then see that I made no attack direct or indirect on himself or his new society.' He insisted that the medical profession was changing rapidly. Doctors could not rely on textbooks because they were invariably out of date; the latest ideas were in the medical journals and this was where new and better treatments were reported. Further, it would be friendlier and 'more in keeping with medical etiquette', Walter said, if practitioners with concerns would approach hospital staff directly, rather than make statements to the press. 'A newspaper is not a suitable medium for discussing medical details; otherwise some of Dr Scott's remarks on nerve cells and muscle fibres and degeneration would be open to considerable criticism.'[40]

For a couple of months, the letter pages of the *Herald* went quiet on the subject of infantile paralysis. But it was not to last. In July the newspaper reported Walter's vehement protest at a decision by the Hospital Board to pay the expenses of the Ladies Massage Society. This gave 'official recognition' to a treatment he opposed and put him (and other medical officers) in 'a difficult position'.[41] It was a characteristically tactful intervention but Walter was adamant. The Board could not logically finance a society that followed a line of treatment opposed by its own medical staff.

Conflicts between the elected members of the Hospital Board and the medical profession were surfacing in other ways, too. There had been unrest at the hospital for several years and an earlier enquiry had failed to address underlying problems. Now the Board was calling for the Matron's resignation, but they failed to produce evidence to substantiate claims that Matron Tait had

mismanaged her staff and they did not give her the chance to answer the criticisms. When a first request for a Royal Commission was rejected by the government,[42] Walter threw his weight behind public demands for reform.

The role of the honorary staff needed to be clarified, he argued. In the
absence of a house surgeon, the honorary staff had attended operations
and given anaesthetics; this was not their proper work and although they
had saved the Board hundreds of pounds, their suggestions had been ignored.
Walter had acted as medical superintendent three or four times but received no
credit or thanks for the time he had given the institution; he had been snubbed
and 'felt it keenly'.[43]

At an open meeting in His Majesty's Theatre, in July 1917, presided over by
the Mayor, Walter spoke eloquently and at length about the injustice done to
Matron Tait, the corrosive nature of such prolonged unrest and the failure of
previous attempts to unravel tensions at the hospital. One of his grandsons
would later describe Walter as having huge presence – and to a young boy he
had seemed a little intimidating. Now, in this public meeting, he held the floor
with confidence and was frequently applauded. He recalled one inquiry run by
the hospital board that had lasted fifteen hours from 2 p.m. until 5 a.m. the
following morning:

> I was summoned at 10 p.m. and by then I had gone 48 hours without sleep.
> I was not called into the inquiry until 2 a.m. Was it possible to give the best
> of my brains? I was nearly asleep. So were some members of the Board. One
> was palpably asleep and another was gallantly struggling to keep his eyelids
> open.[44]

Amidst the laughter, Walter won support and the resolution calling for a Royal
Commission into the running of the public hospital was carried unanimously.
At the second time of asking, the government agreed to set up the enquiry.
About six months later, in the middle of January 1918, a formal public hearing eventually got underway. Those who had called for it, including Walter
and Joan, had every reason to feel optimistic. Surely now they would get to
the root of the trouble at the hospital and action would be taken to resolve
it. Witnesses would be summoned; they would have legal representation and
give evidence under oath. Proceedings were to be chaired by a Mr Bishop, a
respected senior magistrate from Christchurch, and at the end, he would report
fully and openly.

For over a week the columns of the *Herald* were packed with detailed, ver-
batim accounts of proceedings. The Gisborne courthouse was crowded. When
Walter was called to give evidence he testified to an entire lack of discipline
at the hospital. When Matron Tait had dismissed one of the nurses, a Miss
Higgins, he believed she had done so with good reason. Nurse Higgins had

been insubordinate on more than one occasion and had been formally warned. In her evidence Matron Tait explained what had happened:

> Early in January last, I saw Miss Higgins, who was then on duty in isolation, talking to one of the nurses from the main building. They were standing close together – only a few feet apart. A little later I went down and questioned Nurse Higgins who said she had never at any time allowed nurses to come nearer than the fence, which is forty feet away, being the stipulated distance for occupants of isolation.[45]

A few days later, the same thing had again occurred and Nurse Higgins had again lied to the Matron. Then when a Sister Baker resigned, Matron Tait was once more obliged to explain her actions:

> Sister's chief grievance was that she could not manage her ward properly because she could not work hours she had not been used to, so I made the hours the same as those worked in the hospital she came from; but still she grumbled and wanted the sisters' hours changed again. [...] She complained that she had not an adequate staff. Her staff is 6 nurses to 10 patients, and as many nurses as she likes to ask for when there is an operation. On one occasion when she said that she was over worked, I suggested to her that she should give up doing her own washing and baking cakes when on duty.[46]

The Matron had sought support for her actions from the Hospital Board but as other witnesses testified, the Board was divided. Though some of its more mischievous members had recently lost their places, there remained one, a Dr J. C. Collins (who had attended the Pakarae incident with Walter) who quickly emerged as the key troublemaker. He was intent on undermining the authority of the Matron (he had been heard to remark that he would get her out within six months) and the hospital's medical superintendent, Dr Bowie, seemed unable to use his authority to resolve the friction. Further, Dr Collins had been actively soliciting grievances from the nurses, encouraging them to go behind their Matron's back, and then bringing their complaints directly to the Board. Though they were bound by hospital regulations not to leak information, Collins had incited them to do so.

What motivated him? Tensions and altercations between hospital boards and medical staff occurred elsewhere too, most notoriously at Auckland Hospital in the 1880s, 1890s and early 1900s.[47] Typically, elected board members had little medical knowledge and the medical profession fought to take charge of clinical matters themselves. But in Gisborne, the strains had originated with a board member who was himself a doctor and a more likely explanation for the unrest is that he and his associates were seeking to advance their private interests.

Gisborne at the time had seven private hospitals, more than twice as many as Napier, which was a similar size. And it had more private hospitals than the larger north island towns of Wanganui and Palmerston North. Tensions between public and private sectors had occurred in other parts of the country, where doctors frequently complained that public hospitals were too ready to admit free of charge those who could afford to pay, so depriving the profession's members of essential income.[48]

In his final report Bishop pointed directly at those with pecuniary interests in local private hospitals.[49] He did not, he said, receive 'very satisfactory evidence' about the financial activities of Gisborne's medical men but he was able to exonerate Walter and two colleagues, none of whom had any gainful involvement in the work of the private hospitals. Suspicion fell heavily on the others, Drs Scott and Collins, both of whom had owned private hospitals in the town for several years. They had even practised at the same hospital and were close colleagues.[50] Where there was 'so much unaccountable unrest in the public hospital', declared Bishop, the likelihood was that some 'interested parties' may be 'keeping it alive for most ignoble motives'. By undermining and discrediting the public hospital, in other words, some doctors were hoping to persuade more of the town's inhabitants to use the private establishments in which the medics themselves had a financial stake.

As for Matron Tait, she was a 'most capable and efficient woman' and she was cleared of any responsibility for the unrest. 'Few women would have had the courage and grit to see it through', said Bishop. Overall, the blame fell on Bowie, the hospital's medical superintendent. His health had collapsed under the strain of the troubles and his efficiency had been impaired. Change was needed, said Bishop; a return to the old order of things was 'unthinkable'.[51]

It was left to the Hospital Board to decide what action to take and Bowie was eventually forced out, despite support from local residents and from fellow medical practitioners, including Walter. The local newspaper scorned the decision as 'cruelly unfair' and doubted if it would quell the unrest.[52] Walter again took over at the hospital as acting medical superintendent. When Dr Ross, the new medical superintendent, was appointed, he quickly came to the conclusion that he needed a house surgeon rather than honorary staff who would only create more work for him. Walter had no option but to give up his post, and the following day Matron Tait also put in her resignation. Clearly, she was not prepared to remain at the hospital without Walter's support. When her resignation letter came before the Hospital Board, one of its members spoke out bitterly: 'I think', he said, 'that the only thing for us to do is to congratulate Dr Collins and his friends on getting rid of the previous matron.'[53]

The Hospital Board meetings continued to be fractious and dysfunctional and the view that Bowie had been made a scapegoat was widespread. The Royal Commission had *not* after all settled affairs at the hospital and Collins remained in place as an elected member of the Board. It was a depressing outcome for those, like Walter, who had fought for reform. Tony afterwards remembered his father's description of Gisborne as 'an immoral place'. Was this a reference to hospital politics? By this time, the First World War was nearing its close. Wounded soldiers needed beds and it might well have rankled in the town that while so many were dying or sustaining life-changing injuries, others appeared to be lining their own pockets.

13
The Politics of Knitting

The outbreak of the First World War initially unites the local community in Gisborne. Walter and Joan immediately commit themselves to the patriotic cause. Walter provides medical training to the National Reserve and Joan leads groups of women making clothing and equipment for soldiers. By September 1918 they have decided to leave the town.

In August 1914, a few weeks after the outbreak of war, Joan addressed a meeting of the Gisborne Women's Patriotic Committee (WPC). Little Belgium, said Joan, had so far borne the brunt of the hostilities:

> Their country has been overrun by the enemy, their towns shattered and thousands of their brave soldiers have been killed. Intense suffering and poverty will follow and it is the duty of all to do what they can to alleviate that suffering.[1]

The women of Gisborne stood ready to help. They had previously organised themselves and Joan now emerged as a passionate patriot. She urged her listeners to find some little luxury they could do without, and to give the money they saved to the Belgian relief effort. Just as she had once sided with London families living in poverty, now she aligned herself with the subjugated peoples of Europe.

She would have known from newspapers that the war divided English socialists. The British Labour Party (founded in 1906, just before she left the country) endorsed the government's stance. Ramsay Macdonald was an exception and his opposition to the war cost him his leadership of the Labour Party. The Independent Labour Party (to which Harold belonged) also actively opposed the war but most in the Fabian Society backed the government, including the controversial and one-time fellow Fabian, H. G. Wells. Harold let his Fabian membership lapse after 1913, suggesting that he might not have supported the majority's view. He would be remembered by his family for his later opposition to all armed conflict.

If Joan had stayed in England, would she, too, have opposed the war? Her feelings about it, like those of many colonists, were intensified by distance and by the knowledge that family and friends left behind in England were now at risk. To most, the war was not a surprise. For some time it had been regarded as inevitable and it was greeted across Europe by a wave of enthusiastic patriotism. In recent decades some had begun openly contemplating the likely outcomes of armed aggression. Fictional scenarios involving the arrival of foreign armies on native soil (a genre of writing known as 'the invasion novel') had stirred public imaginations and anxieties in England since the 1870s.

Joan was far away now from those she had been close to, in a place where there was little fear of invasion, but as she was growing up a short novel called *The Battle of Dorking* had been a popular reading choice. In it, the Surrey Hills (that would become the favourite location for her trysts with Walter) provided the beautiful backdrop for a fictitious tale about a German attack. Like other examples of the genre, this novel was meant to provoke readers into greater awareness of the foreign threat. From a high ridge near Boxhill, the narrator explains the situation he and his fellow volunteers find themselves in:

> Anybody, indeed, might have been struck with the natural advantages of our position; but what, as I remember, most impressed me, was the peaceful beauty of the scene – the little town [Dorking] with the outline of the houses obscured by a blue mist, the massive crispness of the foliage, the outlines of the great trees, lighted up by the sun, and relieved by deep blue shade ... The quiet of the scene was the more impressive because contrasted in the mind with the scenes we expected to follow; and I can remember as if it were yesterday, the sensation of bitter regret that it should now be too late to avert this coming desecration of our country which might so easily have been prevented. A little firmness, a little prevision on the part of our rulers, even a little common sense, and this great calamity would have been rendered utterly impossible. Too late, alas! We were like the foolish virgins in the parable.[2]

In the new dominion of New Zealand, in the opening months of the conflict, there were few cautionary voices. Most in the community reacted with eagerness, even excitement. The early volunteers tended to be young, single and unskilled men; many were unemployed and many others took the chance to desert their wives. The war initially gave the young country a unifying purpose. Britain's 'most loyal dominion' did not need to debate the matter of going to war and in some places whole rugby teams enlisted together.[3] For the colonists to turn their backs on the homeland was unthinkable. Imperial patriotism was strong – and it was reinforced by more self-interested motives, like the country's dependence on British markets.

Yet despite the overwhelming consensus, beneath the surface some different sentiments were simmering. Pacifists and conscientious objectors were amongst those who protested against the decision to declare war and some women's organisations argued for international arbitration instead. Women in the New Zealand Freedom League and the National Peace Council felt that, as bearers of children, with a natural interest in protecting life, they could not endorse the military effort. Units of Maori soldiers joined the New Zealand expeditionary force but although many Maori were keen to enlist and demonstrate loyalty to the British Crown, others were not, particularly in communities still affected by the injustice of nineteenth-century land confiscations.[4]

As the international conflict deepened, and it became clear that the soldiers would not, after all, be 'home by Christmas' as many had believed, both Joan and Walter became embedded in local campaigns. In Gisborne (as elsewhere) some women attempted to humiliate those who did not enlist by posting them white feathers, despite knowing little of their circumstances, but many others just immersed themselves in patriotic work, as Joan did. It was difficult to be a pacifist in New Zealand in 1914 and she would not have wanted to draw attention to herself by standing out against the popular view. But there was nothing half-hearted about her response either. She was unquestionably committed. And Walter, who was appointed Captain in the Medical Corps of the country's National Reserve in 1915, devoted many evenings over the war years to training field ambulance crews and lecturing on fractures, bandaging and first aid. During the day at the public hospital he was treating a growing number of wounded soldiers who had returned from the front. Hospital facilities were under pressure and, as his professional relationships became increasingly strained, most hours of most days were now taken up with war work.

In the early days of the conflict, the women of New Zealand were called upon to help equip the country's expeditionary force. As in the infant welfare movement and the Plunket Society, it was the social elite who led their sisters in the war effort. Lady Liverpool (wife of the Governor[5]) suggested that in every centre a committee of women should receive contributions. Underclothes, flannel shirts, socks, 'housewives' (or sewing kits) and fabric were all needed urgently. Though working-class communities were also actively involved, it was again the wives of prominent local people, like Joan, who were on the committees and took up the leadership roles.

Over the next four years, just as she had in London, Joan organised campaigns, chaired meetings, compiled reports, lobbied officials and wrote letters. She was active in a plethora of organisations in the town. If there were questions about her background that had gone unanswered since her arrival, her new identity as patriot, war worker and fundraiser would have provided the perfect distraction. Spinning from one meeting to the next, feet barely touching the ground and her head crowded with plans, she had little time to relax. She plunged herself into all manner of activities and with her co-workers she rallied support from the inhabitants of Gisborne. She held sewing bees, patriotic carnivals, and with a friend, Annie Rees, set up a club for returning soldiers.

The family home in Wainui Road became an unofficial war depot, piled high like a jumble stall with donations of clothing, bandages, sheets and pillow cases. Locals were invited to call and collect wool and knitting patterns. A church was converted into a workshop and every Friday, the women came. At busy times (when, for instance, they were making the equipment needed for the two New Zealand hospital ships) there were as many as 60 or 70 workers, and some came day after day, staying long into the evenings to ensure the task was finished on time.

Lady Liverpool was soon joined in her war work by Mrs Miria Pomare, a prominent Maori leader, and together they launched the Lady Liverpool's and Mrs Pomare's Maori Soldiers' Fund, specifically to support Maori soldiers on active service. In Gisborne, Heni Materoa Carroll, whose husband James Carroll had won the Eastern Maori parliamentary seat in 1887,[6] led the fundraising for Maori soldiers. Known to local Pakeha as 'The Lady' (James was the first Maori to be knighted in 1911) she joined Joan on the WPC and was there with her to farewell each contingent of soldiers when they left Gisborne for the front, to pin patriotic badges on their coats and distribute the kits the women had made for them.

On one occasion, in December 1914, with the war just a few months old, Lord and Lady Liverpool visited Gisborne. Joan was present at the first meeting in the town called to organise the occasion and she suggested a promenade concert should be held one evening to boost the Women's Patriotic Fund. The WPC saw to it that the concert was a success – but curiously, Joan seems to have been absent that evening. In newspaper reports, she was not mentioned amongst those receiving the viceregal couple, nor as part of the organising committee.[7] She was named (as Vice-President of the Poverty Bay Women's Patriotic Fund) in a souvenir programme that Lady Liverpool later put in her scrapbook and she may have attended a private dinner for the couple which the Mayor organised.[8] But more likely, she felt it prudent to avoid the spotlight altogether since the activities of eminent English people (like the Liverpools) attracted attention and were widely reported.

Did Joan's associates wonder about the reason for her absence from such an important occasion? Did she feign a minor illness, perhaps? It is impossible to know but she reappeared in public immediately after the Liverpools' departure and continued her patriotic activities with as much intensity as before.

For a country with a population of just over one million, the scale of unpaid war work in New Zealand was huge. Women all over the Empire were knitting, of course, and making clothing to send to soldiers, but New Zealand women had fewer outlets for their patriotic energies than their counterparts in Britain. There were no heavy munitions factories where they might have been employed, and little industry. So they devoted themselves to fundraising (during the war nearly £5 million was raised by about 900 women's groups) and to making equipment for the troops. They knitted on an industrial scale. Hundreds of thousands of socks were knitted by hand, each pair taking about ten hours. In August 1916 alone, over 130,000 knitted items were produced, including balaclavas, bandages, scarves, gloves and face cloths, as well as socks.[9] And parcels (containing tobacco, chocolates, pickles, books and extra clothing) were dispatched to soldiers serving abroad at the rate of 24,000 a month.

Their productivity was extraordinary but the women came to share much else besides the sewing and knitting. A day or so before war was declared, Joan was speaking to a meeting of Gisborne women about their success in cleaning up the town and improving the local beaches: 'I think if you join in some work', she said, 'you get to know people very much more truly than by just meeting them socially. No doubt the bond between us by the work of this committee is a very true one.'[10] She spoke sincerely, in all likelihood drawing on her experience of the industrial women's movement in London. Patriotic work, too, would bring women together, all over the Empire. They would come to know each other through their work, and at difficult times they drew close, caring for and comforting other women as much as the soldiers themselves.

Knitting circles became 'a kind of surrogate family'[11] and songs and poems were shared, in a cultural celebration of the community's cohesiveness. Homespun, like the socks and balaclavas, the poems appeared in local newspapers in many parts of the Empire. Usually written from a woman's perspective, they were staunchly patriotic whilst giving voice to feelings of fear, loss and hope. Sometimes verses found their way to New Zealand in parcels sent home by soldiers, and the families then forwarded them to their local newspapers. At some point early in the war, Lady Liverpool cut out a poem called 'Socks', and pasted it into her scrapbook.[12] Written by Jessie Pope, it begins:

> Shining pins that dart and click
> In the fireside's sheltering peace,
> Check the thoughts that cluster, thick –
> *20 plain and then decrease.*
>
> He was brave – well, so was I –
> Keen and merry, but his lip
> Quivered when he said 'Good-bye' –
> *Purl the seam-stitch, purl and slip.*

Pope's verse was decidedly pro-war (in contrast to the work of Siegfried Sassoon and Wilfred Owen, for example) but it reflected the views of many on the home front in those early days of the conflict. And she was one of the few English war poets to recognise the bravery of New Zealand and Australian troops which she went on to celebrate in poems with an ANZAC[13] theme.

Most of the people of Gisborne were united in their support for the soldiers, and through her patriotic activities Joan formed a strong bond with at least one other likeminded individual. By 1915 she had become friends with local teacher Annie Lee Rees (also known as Lily or Lil). Annie was the sister of Rosemary (who had acted with Walter in the play *In Honour Bound*), and the eldest daughter of William Lee Rees, a prominent lawyer and liberal politician.[14] By the time she met Joan, Annie was a trained lawyer herself and she worked in her father's law firm until his death in 1912. Then, in September that year, she opened the new Cook County College for Girls (shown in Illustration 13.1) in Russell Street, Gisborne, and Joan soon became a supporter.

In 1917, by which time the school had trebled in size, a 'Mrs Reeve's Prize' was being awarded for 'helpfulness and unselfishness'. A kindergarten was now attached and Joan's daughter, Renée, was a pupil there. Aged just five, she won a prize for 'scripture and phonetics'.[15] Girls' education was still a cause that was

Illustration 13.1 Pupils of Cook County College for Girls in front of their school in about 1916. Recent alterations to the house on the right had included the extension of the sleeping balcony for borders

close to Joan's heart but she and Annie Rees also joined forces to support the town's patriotic activities. Joan was at an initial meeting where Rees proposed the setting up of a local soldier's club. Here, men who were on leave or waiting to be called up, or those who had returned to convalesce, could entertain themselves and their friends. Despite some scepticism amongst those present, particularly about how such a scheme would be funded, Joan backed her friend's proposal and urged a six-month trial. A year later, 374 soldiers had registered as members and the club was almost self-supporting.

Just as successful was the Poverty Bay Women's National Reserve (WNR) where again the two women collaborated. Nationally this organisation, at first part of the men's National Reserve, was treated with disdain by officialdom. But some local branches flourished and by 1915, with about 260 members, the one in Gisborne was reckoned to be the strongest in the country.[16] Joan and Annie Rees were both on the organisation's local committee. The women intended to follow the example of their sisters in England and equip themselves to take up vacancies left by men going to the front.[17] They would also prepare to help with nursing the sick and injured, and an initial first-aid course in the town was so oversubscribed that a new venue had to be found. Along with other local doctors Walter gave medical lectures to the WNR, as he did to the field ambulance workers in the men's National Reserve, and talks for women on commercial subjects were also suggested so that they might be prepared if necessary to fill office positions.[18]

New Zealand women were, however, never called upon to do this in significant numbers. Despite questions in parliament and an official recommendation in 1917 to use female labour, as well as exhortations to the women themselves to offer their services, there was no direction of their employment in New Zealand during the war. Fundraising remained the focus of their activities and as it continued in earnest, New Zealand's youngsters were also co-opted. Monday, 10 July 1916, was set aside as a day for the Empire's children to make a special effort to help Belgian children, and at a festival in Gisborne, Annie Rees' pupils from the Cook County College for Girls organised stalls selling produce, and took part in a musical recital.

Even in a country physically far from the front line, children were affected by the wartime attitudes and activities of the adults around them. Elementary schooling was compulsory from the beginning of the twentieth century for both Maori and Pakeha, and most youngsters between the ages of five and thirteen took part in numerous fundraising activities such as carnivals, penny trails, and concerts – and at home, on street corners, in playgrounds and classrooms, both boys and girls knitted constantly.

The Reeve children would remember their Gisborne years, their parents' interminable talk about the hospital all tangled up with the endless knitting. By 1917, aged about nine, Tony and Colin were attending the newly built

CHILDREN'S ACTIVE WORK TO RELIEVE DISTRESS IN BELGIUM AND PROVIDE COMFORTS FOR OUR HERO SOLDIERS: BOYS AND GIRLS SEWING AND KNITTING IN THE CITY SCHOOLS OF AUCKLAND.

Illustration 13.2 Boys and girls sewing and knitting in the city schools of Auckland. This photograph was published by the *Auckland Weekly News* on 15 July 1915

Cheltenham House School in Kaiti where approximately thirty pupils were enrolled. Earlier, they had attended kindergarten, possibly with Annie Rees at Cook County College (which routinely took small boys into kindergarten and junior classes). Pupils from outlying rural districts often had to board at schools in the township and both Cook County and Cheltenham House had accommodation for this purpose. But it could be a lonely time for children and the war sharpened their sense of isolation. In the novel *The Young Have Secrets*, written by James Courage and set in New Zealand during the First World War, a young boy who is boarding away from home receives a letter from his mother:

> I have been helping with Red Cross meetings in the village hall once a week, knitting scarves and rolling bandages to be sent to headquarters in London. The whole Empire must help against these revolting Germans and their beastly Kaiser. Some of us are also collecting clothes for the Belgian refugees who reached England. I have given an old coat of yours, with five brass buttons you had as a little boy, do you remember? Sailor buttons.[19]

Apart from spare clothing, the Reeve children also donated pocket money and took part in fancy dress parades organised by their mother and her co-workers. If they had been a few years older, at school the twins would have found themselves undertaking compulsory military training.[20] They would, in any case, have learnt military-style physical drills which were widely taught to both boys and girls in primary schools. In *The Young Have Secrets*, the protagonist and his school fellows are instructed by a retired major:

> 'Now we'll try marching. A soldier, a British soldier, doesn't walk, he marches. The Empire is his parade ground, he thinks of his king and country.' The words came out in a harsh chant from beneath the moustache: a droplet of spittle caught the afternoon sun on his chin. 'Right. Listen for the command. Left-turn! You, that boy with the crooked tie, don't you know left from right?'[21]

As well as the discipline of the parade ground, New Zealand children learnt songs of the Empire, like 'Pack up Your Troubles in Your Old Kit Bag' and 'It's a Long Way to Tipperary'. And even in remote areas, in small country schools, student fundraising was significant and reported with pride in local newspapers. Usually, the children donated their earnings to a named cause, such as the Belgian Relief Fund, the London Air Raid Fund, the French Orphans' Fund, or the Plum Pudding Fund for Soldiers.

Apart from the fundraising itself, hours of planning, organising, sorting and packing were involved in the war effort – and in Gisborne, Joan led much of the strategic work. Each local committee prepared the clothing and kit for their own soldiers, and by March 1915 confusion had arisen about what exactly was needed. Some equipment for the Gisborne contingent was unused and had been returned. The Defence authorities asked the local patriotic committee to adjust their contributions – and then came word that the recruits in the training camp did not now have enough clothing and bedding and were shivering from the cold.[22] The women were frustrated, angry and demoralised. They had responded wholeheartedly to the appeal from Lady Liverpool and spent long hours preparing the items she asked for. Now much of their time, money and hard work appeared to have been wasted.

Joan was determined to get to the bottom of the muddle. With the Mayoress, a Mrs Sherratt, she took a steamer south and set about making enquiries in person. In Wellington the women interviewed Defence officials, the military commanders, and then the soldiers themselves at their camp in Trentham. They inspected the stores and the catering and obtained copies of official orders and invoices.[23] What had happened to the equipment sent from Gisborne? Which items did the men really need and what was already being supplied by the Defence department? How could waste be avoided? Joan might have been back in London, carrying out her investigations for the Women's Industrial Council, trying to ascertain whether there were jobs for working-class girls in factories or whether employers could be persuaded to help with their training. The problem here was quite different, but her methods just as rigorous and, as in London, they bore fruit.

When she and the Mayoress returned to Gisborne, they submitted an 'exhaustive report' to a special meeting of the Ladies Patriotic Committee.[24] The Gisborne soldiers' kit was the best that had been supplied but the government was now providing the men with good quality clothing at a far lower cost than the women could manage. Joan's recommendation that the Gisborne committee should now focus on making just two items, the 'housewives' and the balaclavas, neither of which was provided satisfactorily by the Defence authorities, was accepted. They would also, they decided, make clothes for the destitute Belgians.

Joan did not spare the country's government in her report. They set up the patriotic committees, she said, and asked them to supply the men's clothing. Yet they never took the trouble to give guidance about the quality and warmth of the clothes needed, provided no sample kits and appointed no one to visit the centres and advise. If they had, an enormous amount of money, both public and private, would have been saved. Gone is the diplomacy and calm of her earlier reports into industrial training for girls. It was not just money that had been wasted. As Bruce Scates observes in his article about war work in Australia, it 'was much more than a tiresome tally of socks, balaclavas and pyjamas'.[25] For hours and hours the women had laboured, putting their hearts into their work, focusing daily on the comfort of others, enclosing thoughtful little gifts or mementos for the soldiers, sitting alongside those who grieved or caring for those in pain. As far as members of the Gisborne Women's Patriotic Committee were concerned, lives were at stake and official incompetence was hard to stomach.

Even worse, of course, was incompetence on the battlefield. Joan would not have known that far away in Turkey the Battle for Gallipoli was about to begin, nor that it would prove such a costly failure for the Allies.[26] Eventually, about a third of the New Zealanders taking part were killed, and overall the casualty rate reached an extraordinary 88 per cent. In a context where organisation was essential to survival, the New Zealand soldiers now became 'deeply suspicious' of the British high command and a new nationalism began to emerge.[27]

As the war dragged on many New Zealanders became disenchanted. In 1916, as the casualty rate rose and voluntary recruitment started to break down, conscription was introduced. Enthusiasm waned and social tensions grew. There were strikes and large anti-conscription conferences; gone was that early sense of unity. And in Europe, Gallipoli was followed by an even worse disaster at Passchendaele. In October 1917 the New Zealand infantry were left 'struggling in a sea of mud and barbed wire', and within a few hours German machine-gunners had killed more than 600 and wounded 2,000.[28] Through the last years of the war, although weighed down by the health care controversy, Walter continued to train field ambulance crews for the National Reserve in his spare time. It brought him pleasure, he said, to work with 'such an attentive and painstaking class'. It would have been welcome relief too, from his duties at the dispute-torn Cook Hospital. For their part, the ambulance men stressed their appreciation; Walter, they said, 'had endeared himself to them all'.[29]

Joan's community work continued, too. She saw the Plunket Society as no less patriotic than the sewing bees and fancy dress carnivals, and in August 1918 she was still suggesting novel ways to raise money for the war effort. On a morning when heavy snow had fallen in the district, she suggested a 'snow ball tea' for Red Cross Day. Each woman would hold a tea for eight invited guests and then they would all become hostesses in their turn, inviting another eight would-be hostesses. If each guest brought one shilling, funds would soon accumulate and the movement would quickly become known.[30]

Joan's energy showed no signs of flagging, but at home she and Walter were by now discussing a new future for the family. Three weeks later, the house in Wainui Road was on the market and Walter had been called up. He had already volunteered for active service twice, but as a married man with three children his application had been refused. This time he was accepted and he announced that he would go into military camp in November.[31]

Illustration 13.3   The Reeve family with friends in 1918, at Waihuka, near Gisborne. In the back row, Walter is second and Joan fourth from the left. Their three children standing in front, from left to right, are Tony, Colin and Renée (on the plinth)

Soon the news that both Walter and Joan were leaving the district led to more warm tributes from the ambulance crews of the National Reserve and Joan's resignation from the Plunket Society which was accepted with 'deep regret'.[32] Then, in early October, came official recognition of her contribution to the war effort. In the first year that women achieved regular access to honours in New Zealand, Joan was awarded an MBE.[33] Local people were immensely proud. The only other Gisborne women to feature in the awards' lists that year were both wives of dignitaries; Lady Carroll (wife of the MP and Joan's colleague on the WPC) received an OBE and Mrs Sherrat (the Gisborne Mayoress who had accompanied Joan on the fact-finding trip to Wellington), received an MBE.[34]

When Joan came to resign from the Soldiers' Club shortly afterwards, committee members made 'eulogistic reference' to her services.[35] A few days later, all the couple's 'high class' furniture (including a piano, two Queen Anne chairs, and several oil paintings) and Walter's two 'first-class' motors, were offered for sale and on 3 November, the family boarded a coastal steamer and headed south. About a week after their departure the war came to an end. And about nine months after that, as the Women's Patriotic Committee in Gisborne was winding up, they acknowledged how much they owed to Joan: 'To her organising ability is due the efficiency and smoothness by which our work has been characterised. We are glad to think that she was able to remain with us practically to the end.'[36]

But with the end of the war came a grim reckoning. The small dominion had made a massive sacrifice. Out of a population of about one million, a total of 120,000 enlisted and about 18,500 died in or because of the war. About 41,000 were wounded[37] and the overall casualty rate was 59 per cent.[38] Though the country had not suffered a loss in prosperity, nearly every citizen knew or was related to someone who had been killed. As Tom Brooking writes in his history of New Zealand: 'The numerous lonely memorials that litter the country and the appallingly long lists of names beneath the columns on school gates and church walls reflect the sadness felt by everyone.'[39]

And despite their tireless efforts at home, the women's gains were slight. Some middle-class women had had the chance to escape the confines of their homes for the first time, but women had not proved vital in filling the employment gap left by soldiers, and the war did not see a massive increase in their employment or a permanent change in their economic status. Both Pakeha and Maori women had already won the vote so did not have to prove themselves 'worthy' of suffrage. They did experience a growth in self-confidence, however, and their expectations changed, says Melanie Nolan, in her study of gender and welfare in the First World War. Previously reliant on their male breadwinners, by the end of the war women were ready to demand more from the state.[40]

As for Joan herself, in all probability she was exhausted. When the family planned the move south to Hawkes Bay, the outcome of the war was still far from certain and she expected Walter shortly to leave them to go to the front. He had already purchased his uniform and kit, including boots, puttees and trench coat. A little anxiously perhaps, he had asked General Henderson (who was in charge of medical services) whether he should travel to camp in his uniform or whether he should be in mufti. And who would give him orders to proceed? After working in civil practice, army discipline was clearly going to require some adjustments. Walter was due to report at Awapuni Military Camp in Palmerston North on 15 November, to prepare for service abroad, but soon it became plain he was needed closer to home. An influenza epidemic of near biblical proportions was sweeping the country. In response to an urgent telegram from the medical authorities, he went instead to Featherston Military Camp, near Wellington, and arrived just two days before the Armistice was signed.[41]

Brought from Europe by the returning troops, influenza was particularly virulent within institutions where large numbers of people were living close together. Medical practitioners were desperately needed in the camps but about a third of the country's doctors were still overseas. Those who were available to deal with the epidemic were at considerable risk of infection themselves and nowhere more so than in the camps. This was one of the most dangerous places you could be.[42]

Walter was by now no stranger to epidemics. Most recently, of course, he had witnessed the 1916 polio outbreak that had affected the children of Gisborne, but he would also have remembered how quickly several contagious diseases had spread through the Church Missionary Society Home some forty years before, when he was a boy. Bacteriology was leading to the development of new vaccines, for diseases such as typhoid and diphtheria, but influenza was still often fatal. As the disease spread through local communities, carers (including volunteers and family) were urged to isolate the sufferers. Though recruitment to the Expeditionary Force had now ceased and it was plain Walter was not going to be sent overseas, Joan had no less to fear from his trip to Wellington that year.

14
Landfall

Joan and Walter take the family to live in the tiny village of Havelock North, in Hawkes Bay. Walter soon leaves for the Featherston military camp, near Wellington, where he treats victims of the influenza epidemic. Meanwhile the children settle into local schools. The 1920s bring mixed fortunes; the children do well academically and Walter's medical practice thrives, but Joan receives a troubling diagnosis.

The Reeves arrived in the small rural village of Havelock North in early November 1918. Walter had barely a week to settle the family into their new home before he headed south to the military camp at Featherston. The number of servicemen there who were ill with influenza had been climbing steadily since late October. Then, on the night of 6 November, there was a fierce storm that brought down camp buildings, flattened tents and forced the men to crowd together, facilitating the spread of the disease. Walter arrived just before it reached its peak. By now, almost 2,500 men were ill. The authorities converted several camp buildings for use as makeshift hospital wards and medical staff proceeded to treat the sick men with sodium salicylate and shots of alcohol. More than 170 men died in the Featherston Camp but later, the Principal Medical Officer claimed that the use of alcohol had saved many lives.[1]

When the number of sufferers in the camp had declined, Walter was sent to the nearby town of Carterton. Here, people were in need of medical assistance because the local doctor had himself succumbed to the virus. The inevitable happened. Away from the camp facilities, the isolation wards (and perhaps the whisky), Walter himself contracted influenza. The death rate amongst medical staff across the country was almost as high as that for soldiers in the military camps. As Walter would have known, the danger was not so much the disease itself but the possibility that it could lead to pneumonia which was the real killer. Patients developing this secondary infection often found their skin and nails turning black as the body was deprived of oxygen.

Walter was ill for about a fortnight and he delayed returning to the family in Hawkes Bay until two days before Christmas. They would have been relieved to see him. By this time a Health Committee had been set up in Havelock North to combat the influenza epidemic in the village and surrounding district. In the chair was local doctor, Robert Felkin. Walter would soon take over Dr Felkin's medical practice, but for now the two men worked together, treating those who had contracted the disease and trying to prevent its spread. They appeared to have some success. Not one fatality occurred, although 300 deaths were recorded elsewhere in Hawkes Bay.[2] By the end of the year the epidemic was over, but most children would remember the influenza of 1918 more vividly than the war itself because the virus came into their homes and affected adults close to them. Across New Zealand, more than 8,500 people died; influenza had created more orphans than Gallipoli.[3]

In early 1919 Walter was still under military orders. After a brief spell at Trentham Military Camp (a smaller camp that was also near Wellington) he was sent to Auckland. From February to August he was in charge of orthopaedics at what would soon be called the Auckland Military Convalescent Home. In his first month he carried out over thirty operations on ex-servicemen. By April he was seeking to be relieved of his duties. By June, when his request had again been refused, he was becoming more insistent. He was anxious to return to his family. His army pay was insufficient to cover the cost of his three children's schooling, as well as his life insurance. At 43 years of age he was junior in rank to colleagues who were younger and had less medical experience: 'This position is full of difficulties', he wrote, 'and not very comfortable. When I joined up the country was actively at war so the question of Home, Pay and Rank did not count for much. One feels differently about them now.' In addition, he was needed in Havelock North by Dr Felkin who was elderly, unwell and under 'excessive strain'. Felkin himself also wrote to the military authorities, hoping to be forgiven if he was 'overstepping the bounds', but urging them to accept Walter's application for demobilisation.[4]

The wheels of army bureaucracy turned slowly. This would have been no surprise to Walter who was already engaged in a protracted dispute over allowances. Before leaving Gisborne, in anticipation of being sent abroad, he had purchased the clothing and equipment he was told he would need, but then discovered that because he had not gone to the front after all, he was to be reimbursed at a reduced rate. In fact, he did not receive any allowance at all until August 1919, the month he was finally able to return to civilian life and to the family in Havelock North.

At this time, Joan and the children were living in Campbell Street, but within a few years the family had moved into a colonial-style wooden bungalow on the corner of Duart and Te Mata Roads. Built around the turn of the century by a local family, the house (Illustration 14.1) was typical of its era.[5] In a nostalgic gesture, the full significance of which would have eluded their new neighbours, Walter and Joan named it 'Ranmore'. At first, it was painted a fashionable, deep red. Later, Walter made extensive alterations to it.

Walter now took over Dr Felkin's medical practice. Felkin was a prominent if controversial figure, always cloaked in a little mystery, and he had long nurtured interests in the occult. Some locals even suspected it was his 'magical powers' that had brought the village safely through the influenza epidemic. Felkin himself was gracious enough to acknowledge Walter's assistance in the matter[6] but he wanted now to devote himself to a local spiritual community known as 'Havelock Work'.[7]

This organisation had been established in 1908 and had its roots in the Hermetic Order of the Golden Dawn, one of many occult groups that flourished in nineteenth-century Europe. Initially, 'Havelock Work' members did little more than gather to hear readings from Shakespeare and Dickens. Soon, however, there were classes in drama and carving, medieval processions, concerts and in 1912, a Shakespearean Pageant. Behind the festivities lay a serious purpose and a hope that by cultivating a feeling for things that were 'beautiful and true'[8] the way could be found to those secrets and mysteries that had been promised to the early Christians.

Illustration 14.1 'Ranmore', in Duart Road, Havelock North. In the 1920s Walter set up his consulting rooms to the right of the front door. The house was moved to a new location in 2010

Members combined a kind of high-minded spirituality with reformist zeal and social concern. They observed the solstices and the equinoxes, and adopted bizarre rituals and ceremonies to initiate novices who were led, robed and blindfolded, down into a painted basement temple that remains to this today. Here, you might spend a whole night entombed in a small, seven-sided room, contemplating the meanings of the signs and symbols that filled each wall from top to bottom. Though the group conducted its activities in secret, its membership was believed to include nearly all the village's most eminent and respectable citizens who were always able to thwart any official enquiry into the organisation. By 1926 there were about a hundred in the inner circle, and most were devout Anglicans.

But Walter and Grace were not amongst them. 'Havelock Work' had its headquarters not far from Ranmore, in a house called 'Whare Ra' ('House of the Sun'),[9] and initially the organisation may have sounded to the couple uncomfortably like the now defunct 'Fellowship of the New Life'. Just as Harold had argued in the 1890s for a rejection of materialism, so 'Havelock Work' members, too, sought uncluttered lives and time spent in harmony with both nature and the divine. There were clear differences between the organisations, however, and the New Zealand group's secrecy and fascination with magic set it well apart from the ethical socialism of late Victorian England.

As well, now that the influenza epidemic was over, Walter found he had no time for Felkin's more unorthodox medical practices. As in Gisborne, if he was needed Walter would be out in all weathers and at all hours, but Felkin had appeared to be selective about those he treated. As well, he practised some alternative therapies, notably one in which children were exposed to ridicule by being made to wear stockings of different and mixed colours. The practice had unproven benefits and, unsurprisingly, it made the wearer miserable. Although the ever tactful Walter remained diplomatic in public, at home he was less restrained and to his family he voiced disapproval of some of his predecessor's methods.

Though their parents did not associate with the Felkinites, Tony and Colin and their sister, Renée, may well have been curious about the eccentric goings-on a short distance away at Whare Ra. Long after Robert Felkin himself had died in 1926, his elderly widow continued to hold Christmas parties there for the village children. Many years later, Tony's daughter would recall that on these occasions, there was always a huge Christmas tree in a darkened room, and it was decorated with real candles. A short story by New Zealand writer, Barbara Anderson, who also grew up in Hawkes Bay, is based on her childhood recollections of these strange festivities. In 'Poojah', the children are pressed to kneel in front of a nativity scene. 'We were meant to make the sign of the cross', says the narrator, then 'strike the censer three times against its base – and say in clear tones, "Happy Birthday, Baby Jesus".' Once rituals were concluded, the children were given tea 'in another shadowed room. Why was it called the House of Sunshine, when the sun was so rigorously excluded?'[10]

And why, in the early twentieth century, did 'Havelock Work' flourish here? How did this tiny Hawkes Bay village become the setting for such an extraordinary story? Certainly, after the carnage of the war, spiritualism was in vogue. Interest in theosophy and utopianism was a feature of intellectual life in Europe. But the Havelock North group had become established well before the conflict. Robert Ellwood, the American academic, suggests that its very isolation may have been a factor.[11]

Havelock North had originally been founded by the government in the 1860s to provide land for small farmers and working-class settlers, but most of the plots were bought by speculators and wealthy pastoralists. The small farms were unable to develop and the growth of Havelock North, like that of other towns in the region, was restricted by the surrounding pastoral stations, at least until the late 1890s. When Walter and Joan arrived in 1918, there were barely a thousand inhabitants but they included the elite of the region. In contrast to Gisborne, Havelock North was prosperous. This isolated and homogenous community of upper-class settlers now attracted the educated, the creative and the affluent. Private schools were established for children of wealthy landowners and soon, Havelock North became the cultural centre of Hawkes Bay. Some prominent families were Quakers who had a history of religious tolerance, and others were liberal Anglicans, with an interest in mysticism. It was fertile ground for the charismatic Felkin and his enterprise caught the mood of the new century. Here, there could be a fresh beginning and Victorian stuffiness would be swept away.[12]

But Joan and Walter, both now in their forties, had more earthly concerns. Once the war was over, the epidemic had passed and Walter was back home, family life could resume. Despite the sorrow and loss experienced by so many around them, it must have felt like a new beginning. Several schools in Havelock North had good reputations and Tony and Colin were already attending the private Heretaunga School (later renamed Hereworth). Renée was at St George's primary school nearby. In 1926, aged about fourteen, she began her secondary education at Woodford House, a local boarding school for girls. By that time her brothers were boarding at Wanganui Collegiate School, some distance away on the west coast of the North Island.[13]

The children did well at their schools; the twins were both House Prefects, and Renée won several prizes for her academic work, particularly her poetry. Walter's new patients were less likely to become bankrupt and default on their payments than those in Gisborne had been, and he was free at last from worry about the local hospital and the activities of a few self-interested colleagues. He still enjoyed driving fast (he was adept at talking his way out of speeding fines) and he also found time now for golf (he was the Havelock North club's first president); the Boy Scout troop (he was a member of the founding committee); and swimming (he helped build the local village pool that would last for nearly forty years). Joan was less prominent. As the wife of the local doctor, she took an interest in community and civic affairs, but after her strenuous work in support of the war effort, she now adopted a less conspicuous role. We catch a brief glimpse of her in 1925, when she was leading the first Girl Guide Company in the village and, until the group became too big for a private house, she hosted their gatherings at Ranmore. She also became the first chairwoman of the Havelock Women's Institute, though no records of this organisation survive.

Beyond the village, in the 1920s, the national mood was changing. The decade brought mixed economic fortunes for many New Zealanders and it was a sombre time. Returning soldiers, far from receiving a heroes' welcome, were met with little sympathy and even less understanding of their medical and psychological needs. Some who were lucky enough to obtain land under the returned soldiers' settlement scheme found they were in possession of plots that were overvalued and overgrown. Many had walked off them in despair by 1925.

Social divisions began to sharpen. In a country so far devoid of the kind of sectarianism that had characterised Britain, the United States and Australia,

Illustration 14.2   A street in Havelock North, Hawke's Bay, in about 1925

religious bigotry now surfaced. Prohibitionists intensified their campaigns and several times in the 1920s, they almost succeeded. There was a frenzy of puritanical disapproval about such immoral activities as modern dancing and the society's cultural conservatism led to the departure of several writers, including Jane Mander, who felt unsupported. By the late 1920s depression loomed. The country's material progress had been undeniable. Prices for vital farm exports had fluctuated but technological advances had brought more cars to the roads, electric stoves into homes and a golden age of rail travel. Both Pakeha and Maori populations increased. By 1926, sixty-one per cent of New Zealanders owned their own homes, probably more than in any other country in the world.[14] Even so, by 1930 the economy was in severe crisis and the whole point of emigrating so far from Britain seemed to have been lost. For many, the utopian dream was in bits.

There were some who still believed, who cherished the original nineteenth-century myth, and over the coming decades, it remained, says Ellwood, at the back of the collective mind.[15] Throughout the twentieth century it surfaced time and again, in various socialist and welfare reforms as, for example, in the idealism of the Plunket Society that Joan and Walter had worked so hard to promote in Gisborne.

And the Reeve children remembered happy times in Hawkes Bay. On one occasion, the twins built a flying fox by running a wire from the top of the chimney at Ranmore down to a tree in the garden, and every Christmas the family spent their summer holiday camping and sailing at Horseshoe Bend, on the Tukituki River. Here, surrounded by softly folded, green hills, there may have been little to remind Joan and Walter of earlier sailing trips off the coast of England, but they were always happy to be on the water, and never more so than in this remote and beautiful place. One year, Tony took his mother out in a boat he had made himself and it capsized. Joan was fully dressed and as she surfaced, Tony heard a gurgling sound under the water. He was puzzled, perhaps worried, until he realised it was the sound of his mother laughing.

A more unsettling memory was of whispered conversations which the children overheard. News had been received that someone had come looking for Joan, presumably from England. Who could it have been? The couple believed they might have to leave Hawkes Bay to avoid detection and Walter even travelled south to explore alternative employment. Nothing came of the rumour but living with a constant fear of discovery would have taken its toll. Like a swan, Joan appeared serene but underneath the water she must have been paddling furiously. Such a prolonged act of deception, the extended separation from her family and friends in England, the guilt she may have felt at betraying the trust of her new associates may all have had some psychological consequence. As the only person who was party to her secret, Walter would have played a pivotal role. He was her confidante, and perhaps he knew how much effort it took; perhaps at times she kept it even from him.

When did they first put a name to her numbness and her loss of balance? When did they start to realise the blurred vision was not going away? Multiple sclerosis is stealthy in its approach. No scientific evidence links its onset with stress but it is more prevalent amongst women and in countries further from the equator. Typically, the symptoms come and go at first, as the body slowly shuts up shop. Though it is one of the most common diseases of the nervous system, even today diagnosis can be difficult.[16] And in Joan's case, it may have been especially so. Walter would not have wanted to see that the symptoms were anything more than fatigue. In the village, he did not have a circle of trusted colleagues he could call on to confirm or deny his fears. A hospital was planned in nearby Hastings (as a memorial to fallen soldiers) but it did not open until 1928 – the same year that Joan came to understand the reality of her situation, drew up her last will and testament and signed it, printing her name carefully and with obvious difficulty.

If Walter had wanted to keep Joan's illness from the children, for a while at least, it may have been relatively easy. The most likely date for the confirmed diagnosis is 1926, a year when all the children were away at boarding school. Renée was the nearest since Woodford House was in Havelock North, but she was allowed home only on two or three Sundays each term. Tony and Colin did not return home during term time. To reach their school on the west coast they had to travel by train from Hastings, passing through the Manawatu Gorge between two mountain ranges, on a journey that took a whole day. The school must have seemed particularly far away to Joan and Walter in 1926 when, at the beginning of a new term, Colin developed appendicitis and was operated on, the staff said, 'just in time'.[17]

Joan died on 11 December 1929. Walter was with her. She had been ill for at least three years and towards the end she suffered badly. By now a nurse was in constant attendance. In reply to a letter of condolence which he had received from a family friend, Walter spoke of Joan's 'weary days' as her illness had progressed. 'None of us could wish her back however much we miss her', he said.[18] Colin and Renée were home for the long summer holiday but Tony had left to study medicine in Edinburgh the year before. He knew his mother was experiencing some difficulties but it appears he was not fully aware of the nature of her illness before he went away. Once he received word from Walter that she had little time left, he set off on the return voyage to New Zealand but arrived too late to see her. It remained a source of anguish to him for the rest of his life.

After Joan's death, Walter set about completing the necessary formalities. On 13 December he buried Joan privately in the local cemetery and registered her death. The official record is a bizarre mix of fact and fiction. To keep faith with information they had previously supplied on the children's birth certificates, and probably elsewhere too, Walter carried over Joan's invented maiden name of 'Knight'. Now, for the story to be consistent, it became necessary to amend her parents' names. Walter retained their real first names and they became 'James and Elizabeth Knight'. It was a strange gesture, a half-truth that linked Joan with her (real) past, and those she had left behind in England. James' profession was accurately recorded as 'secretary', but Joan's birthplace was another lie. She was born not in Highgate in London, but in Hackney. And she was not 54 when she died, as stated on the official record, but actually 57. Further, if her age as entered on the *Manuka*'s passenger list in 1907 had been correct, she would now have been only 47. Walter's marriage to Joan, which never happened, was now recorded as having taken place when she was 32, in 'Branston, Yorkshire'. There were several places in England called Branston, but none in Yorkshire. A few years earlier, in his application to join the New Zealand Expeditionary Force, Walter had given the place of his marriage as 'Brixton, Scarborough' – which did not exist either. But the children's names and ages are correct, as well as other instantly verifiable details such as the family's address and the cause of death.

Despite these threads of truth, however, the fictional fabric was now strain-ing at the seams. Perhaps it is easier to lie in part and to build from the truth rather than create a new but completely coherent version of events. Perhaps Walter wanted to anchor Joan's death in some kind of reality at this sad moment in his life. Perhaps in the distress of the occasion, he forgot the details of their earlier efforts at concealment; certainly it must have seemed to matter less. Havelock North was a very long way from some place called Branston no matter which county it was in. In a further effort to preserve her secret beyond the grave, Walter put very few details on Joan's headstone. Noticeably absent is any date or place of birth, any parentage, or details of her bereaved family.

How likely was it that all this subterfuge would raise rather than bury speculation? Walter might have wished to close down any possibility of dis-covery but he need not have worried unduly. The reality was that, in the late nineteenth century at least, people's death records in New Zealand were often incomplete. Between 1876 and 1890, sixty-five per cent of death certificates issued in Wellington lacked biographical details such as place of birth, length of time in New Zealand, father's name or occupation, or mother's first or maiden name. Especially during gold rush days, many migrants had transient and isolated life styles, and they died alone.[19]

And now that Joan herself was dead, what was the worst that could happen? Few people in New Zealand at the time would have had any cause to doubt the couple, let alone the ability to find and challenge the inconsistencies in their story. But reputations were still important. The day after Joan's death a short

Illustration 14.3   The headstone on Joan's grave in the Havelock North cemetery

tribute appeared in a local newspaper. Like Grace Marion Oakeshott, mourned more than twenty years earlier, Joan would be remembered for her personal and professional strengths. Her contributions to the local Guide movement and the Havelock Women's Institute were warmly acknowledged. Her 'lovable nature', ran the article, 'endeared her to a large circle of friends who will sincerely regret her death'.[20] Walter would have wanted to preserve Joan's good standing amongst those who knew them and it was easy now to stay loyal to the fiction. Probably, he had promised Joan it would be so, and nearly thirty years later his own death certificate would again record details of his fictitious marriage, at the age of 30, to one Joan Leslie Knight.

There were people in England, too, who must never be allowed to find them. What had happened to those Grace left behind more than twenty years before? In particular, what had happened to Harold, the abandoned husband? It was only in tracking his story that the truth finally emerged.

15
After Lives

Back in England, Harold remarries and Grace's siblings have successful careers. In New Zealand, after Joan's death, Walter forms a close relationship with a vicar's wife. Eventually, Tony, Colin and Renée learn the truth about their mother.

It was about three weeks after Grace's disappearance that the short news item describing Harold's 'cruel blow' was published in England in *The Croydon Times*. His wife had drowned in France and, the reader was told, there was 'small chance of the body being recovered'. Did the wording of the piece seem odd at the time? Why not say that her body had not *yet* been recovered? No notice of a funeral or memorial service appeared over the next few weeks although tributes and obituaries were published, including the one in *The Times*.

Harold's family was keen for him to remarry and about a year later, he did so. In the marriage register he described himself as a clerk and a widower. He was by now 37 years old and his new wife, Dorothy, was fifteen years his junior. For some time they continued to live in the house in Coulsdon that Harold had shared with Grace. Between 1909 and 1917 they had four children who all eventually learned Grace's secret from their father.

Sometime after the sailing trip in the summer of 1899, as we know, Walter and Grace began an affair and in 1907 Harold was complicit in his wife's disappearance. His children (and later their children) told and retold the story, like 'a family joke'. As always happens in games like Chinese Whispers, the narrative that eventually came to light contained some errors (the couple were said to have gone to South Africa, for example) but there was no denying its underlying veracity.

Harold was not in France with his wife that summer, but in all likelihood it was he who first broke news of her 'drowning' to friends and colleagues. The obituary in *The Times* then appeared, backing up his account, adding weight and authenticity, and the item in *The Croydon Times* eight days later, with its bold insistence on the improbability of finding Grace's body, was a further ruse to throw friends and colleagues off the scent.

In remarrying the following year, Harold knew (as Dorothy did) that he was committing the crime of bigamy for which he could be imprisoned. In common law, if there was no body relatives had to wait seven years before a missing person could be presumed dead.[1] So this was a secret that had to be managed carefully. Did the registrar ask Harold for evidence of his widower status at the time of his second marriage? If so, then both newspaper items would have provided valuable heft.

Despite the illegality of Harold's actions, bigamy in itself was not uncommon and judges were often lenient in such cases. Divorce was known to be prohibitively expensive. If there were no children who might be neglected as the result of a second union, then local communities too, especially working class ones, might turn a blind eye.

In radical middle-class circles, however, amongst the Fabians and members of the women's industrial movement, there was more at stake. H. G. Wells had created anxiety amongst socialists with his advocacy of free love and his own unorthodox living arrangements. And a number of Grace's colleagues on the Women's Industrial Council (including Margaret Macdonald) held traditional beliefs about marriage and women's place in late Victorian society. Even in progressive circles, therefore, a strong attachment to the value of respectability might temper people's actions. Grace and Harold did not want to bring their families into disrepute but it is likely they felt it just as important to avoid discrediting the socialist movement. They had no wish to cause controversy, to jeopardise their own social and professional standing, or that of others. One obvious way to avoid all shame and to keep all reputations intact was for Grace to 'die'.

Why did Harold agree to it? Probably, he knew he had already lost her. By 1905 Grace was employed in her full-time post at the London County Council, and she no longer needed his financial support. She was in love with Walter – and Harold was principled. A life worth living, he had insisted, was one lived in comfort and security, with opportunities for self-realisation and freedom from artificial constraint. Everyone deserved the chance to fulfil themselves and find happiness.[2] His decision to set Grace free was entirely consistent with his political philosophy, though it cost him dear. In the words of his granddaughter, 'Who knows what chagrin in his heart?'[3]

For respectability to be preserved, family members would have to be sworn to secrecy. There is no one alive now who can tell us how they felt about Grace's plan, but they must have agreed to collude with it and to let it be known that she was dead. Within a family, various fictions could be sustained but consistency was vital. In 1907 *The News of the World* regularly carried a column headed

Illustration 15.1   Harold with Dorothy, at the time of their engagement

'Missing Relatives'. Here, a sister, brother, mother, cousin or nephew could advertise for news of their loved one, who had disappeared and last been heard of in Burma, Chicago or Walthamstow. Those placing the advertisements were warned that this column was for genuine cases of missing relatives only and not for runaway husbands and wives, and that they should not send money until they had confirmed that the respondent was definitely the person they were looking for. Potentially, it would have been disastrous if Jessie, Kate or Henry, for example, had advertised for news of their missing sister in this way. So, in 1907, the families had to be told the truth. And there had to be an explanation, or some kind of narrative that could be passed down through the generations. Neither of Grace's sisters married or had children, but Henry's son (Grace's nephew) remembered being told that Grace had disappeared and that they 'did not talk about her'. The Oakeshott family inherited a more romantic account that later inspired Harold's granddaughter to imagine Grace's new life in poems and songs:

> The romance was gaily spoken of by my granny, mum and aunt
> Who were tickled by this story
> Of what must have been
> A bigamous like stunt.[4]

The Cash family appeared to take the matter more seriously. Through the years following Grace's departure from England, they carefully erased her from the official records. Deborah Cohen, in her study of the way Victorian families coped with secrets, describes how numbers of children born with disabilities in the first half of the twentieth century similarly 'disappeared' as their families, anxious to avoid the stigma of abnormality, consigned them to institutions where they lived out their days unvisited and unacknowledged.[5]

Like many of these children, Grace was also systematically deleted from public documents. Her parents moved away from Fanfare Road in Coulsdon soon after Harold's remarriage, but they continued to safeguard their daughter's secret. When James, now living in Blenheim Crescent in Croydon, completed the Census in 1911, and indicated that of his four children only three were still alive, he was not the first person to log an untruth on a Census return. Some people entered their children as 'scholars' to disguise the fact that they were sending underage offspring out to work. Others were inventive about their ages. In 1911 significant numbers of women boycotted the Census altogether out of anger at the government's refusal to grant them the vote. It was the first time the country's citizens had been asked about the number of their children who had died, and the question was intended to meet government concerns about a falling birth rate. James might not have expected it but he was not caught out. Then, the following year, he revoked all his former wills and left his estate to be divided equally between his three children whom he named as Jessie, Kate and Henry.

Did any correspondence pass between Grace and those she left behind? In 1912, when her mother, Elizabeth, died in Croydon did her father write to tell her? And did Walter tell the Cash family when Grace herself died in 1929? Did he tell Harold? The families know only of one letter and that was written by Grace herself to Harold, soon after the twins were born. In it she tells him she is very happy. She has two children, she says, and she always thinks of him with respect. Intriguingly, Harold kept the letter. It was Dorothy who destroyed it, but not until after her husband's death in the 1950s.

Several members of Walter's family were also party to the truth and similarly briefed. Walter himself did not need to 'disappear' or lie on his own account.

On the contrary, since he wished to practise medicine in New Zealand, he needed to preserve an identity that was consistent with his entry in the British Medical Register, and after his arrival in New Zealand he had to show evidence of his medical qualifications. He had to protect the real identity of the woman posing as his wife, of course, and he could not afford to do anything that would undermine his claim to have married a person called Joan Leslie Knight. His decision to live with another man's wife would have been hard to sell to his devout, evangelical parents, but by 1907 his mother, Emily, was dead and it would be some years before he revealed the truth to his father.

Walter's older brother, Herbert, had by now married Anna Carr (the sister of Stanley, Walter's boyhood friend and sailing companion), and the couple were living in South Africa. Their descendants, too, knew 'the family scandal' and talked openly about it. When Walter left England, his older sister, Kate, was still in Croydon. Like Grace's sister, Kate Cash, Kate Reeve was a kindergarten teacher. She was also friends with Grace and Harold – and she, too, eventually became a confidante. In 1906, a few months after her mother's death in Canada, Kate Reeve married Charles Mitchiner and although she died in childbirth just four years later, her husband lived until 1970. In the years after Grace's death, he became a key informant in the story, along with Dr Francis Carr, another of Anna's brothers.[6]

With so many of the extended family in the loop, it is surprising that the secret did not leak to others while Grace was still alive. Of course it might have done, and this may explain the mystery stalker in New Zealand in the 1920s. Now, more than a hundred years later, it looks very much as if the only people we can be sure were genuinely duped were Harold's comrades in the Independent Labour Party, and Grace's sisters in the movement for the industrial progress of women, those who wrote the warm, loving tributes to her, expressed their grief so graphically, and treasured her memory.

If 1907 was the year Grace decided to change her life irrevocably, it was also the year her sisters took decisive steps in their careers, as if perhaps they might have thought that now they, too, could achieve their dreams. When Jessie came back from the educational tour of America, Sydenham Road School had closed for Christmas, and she did not return to it. Instead, in early January 1907, she became the first headmistress of the new Winterbourne Junior Girls' School, in Thornton Heath. Winterbourne had opened a year earlier as a combined boys, girls and infants school, but when Jessie arrived it split into three separate institutions and on a salary of £140 a year she took charge of the girls' school with 370 young pupils. She had brought four staff with her from Sydenham Road and it was a fresh start for them all. She was full of ideas from her American trip.

In 1907 Jessie (Jess to her family) was living in Croydon on her own, but by 1911, now aged 42, she had moved back to the family home. Many single women of her generation who were without independent means were likely to become the mainstay of ageing parents. But Jessie was earning a decent salary and had lived away from the family for some years. Probably her return was motivated by duty and affection rather than her own need for material support. Her mother died shortly afterwards and her father about ten years later, in 1922.

Jessie remained in post at Winterbourne for almost 25 years and she was popular. Much later, her former pupils remembered writing on slates and, during the First World War, sharing their classes with Belgian refugees. By the time Jessie came to retire in 1931, the school was creating 'a very pleasing impression of industry and effort'. In recognition of her 'loyal' and 'enterprising' work, she was formally presented with a McMichael portable wireless set.[7] Broadcasting was still in its infancy in the early 1930s but the McMichael company was already well-known for its range of portable products. Jessie died in 1953 at a Croydon nursing home. The two Winterbourne Junior schools are still unique as the last remaining single-sex, local authority maintained junior schools in the UK.

As for Kate Cash, Grace's second sister (known as Kaye in the family), she too reached a turning point in 1907. Two months after Grace left her clothes on the beach in Brittany, Kate registered at University College London and began her studies for a BA degree. Some decades before, London had become the first university to admit women on equal terms with men and by 1900 they comprised about thirty per cent of its student population.[8] Kate excelled. In 1908 she passed the intermediate arts examination (she took Latin, Logic, English and History) and she won the competitive Derby Scholarship in history, which was worth £50. She was by now renting her own accommodation in St John's Wood, and the extra funds would have been welcome. In 1910 she graduated with a first-class honours degree in philosophy, and also won a further prize, this time in Greek philosophy.[9]

For the next ten or twelve years she lectured in education. She taught trainee kindergarten teachers at a number of schools and colleges in and around London, and then in the autumn of 1923, about a year after her father's death, she returned to the Froebel Educational Institute, where she had completed her own kindergarten training some thirty years earlier. It was now located in Roehampton, in Surrey. She was employed here temporarily and part-time, initially for just 15 shillings an hour, teaching mainly History and Psychology. Then, in 1926, she obtained a full-time post at the FEI, on a salary of £300 a year. Madame Emilie Michaelis, who had so inspired Kate and her fellow students in Croydon, was no longer the Principal but it must nevertheless have seemed the ideal job. The Institute was expanding and in this prestigious home of kindergarten teaching, Kate now had the opportunity to influence other young women, to persuade them to follow Froebel's teachings, as she herself had done.

All seemed to go well at first. Kate was now living in Chelsea, and she received regular salary rises. By April 1929, however, she was ill. Staff records show her absent on sick leave and then in May 1929 she is reported as suffering 'a nerve breakdown'. Her return to work is 'still uncertain'.[10] By the summer term her employer has received notification that she will not be coming back and she disappears from the Institute's records.

Could there have been any connection between Kate's breakdown and her sister's illness in New Zealand? If Walter had by now told the Cash family just how ill Grace was, and we do not know if he did, it might possibly have had an impact on Kate's state of mind. But the sisters had not seen each other for over twenty years so it is just as likely that the timing was coincidental. What we do know is that, happily, Kate recovered. By 1933 she had moved to the small settlement of Headley Down, in Hampshire. Here, she became associated with Dr Elizabeth Wilks, a suffragist and medical practitioner, who had moved from London with her husband, Mark, a teacher. The couple had hit the headlines more than twenty years before, when Wilks had refused to pay her income tax, arguing that women who could not vote should not be liable for it. Eventually Mark Wilks, who refused to pay it on her behalf, was imprisoned. He was released two weeks later and the couple were celebrated by suffrage campaigners for highlighting anomalies in the taxation system and for taking a stand on equal rights.

In Headley Down the Wilks lived in a cottage called 'Openlands', in Furze Vale Road. Concerned that so many in their local community were in overcrowded accommodation and lacked basic amenities, with characteristic energy and zeal they set about finding a solution. By the time Kate arrived they had established the Headley Public Utility Society which built sixteen new cottages. These were then let to local working-class families at affordable rents.

Kate lived a short walk away from Openlands, in the next road. She would have enjoyed spending time with her neighbours, the Wilks. Both the women had a keen interest in social affairs and both were linked with the suffrage movement.[11] When Kate died in 1949, she left Elizabeth Wilks the first refusal on buying her bungalow at a 'reasonable not inflated price'. If Wilks predeceased her, the property was to be offered for sale to the Headley Public Utility Society. Kate's obvious intention was that her cottage, too, was to be used to house local people. Wilks died seven years later and a warm tribute in the local newspaper recalled her 'indomitable will' and her strong views. She had the ability to 'arouse the conscience of others' but enjoyed company, and in the 1930s often sat with friends round a log fire, deep in conversation.[12] These were happier times for Kate. Her cottage 'Solway' no longer exists but, thanks to Elizabeth Wilks, about ten acres of woodland, known as Openlands, is today held in trust for the enjoyment of local people.[13]

By the time Grace left England, her younger brother Henry (known also as Harry or Hal), now a fully qualified electrical engineer, had established his own company. It was a propitious time for the industry and H. J. Cash & Co. quickly achieved a solid reputation. Henry was an active member of the professional body and in 1918 was elected President of the Electrical Contractors' Association.[14] Over the next two years he led the organisation through the tumult that followed the First World War, when the industry lost many members and young apprentices and the sector faced massive reorganisation. He became known for his strongly held opinions and, just as when he was sailing with his friends and family round the English coast, in his professional activities he ran a tight ship. His manner was direct and his presentations and letters to members often caused a stir. But he was witty, too. In a speech at a dinner to celebrate peace in 1918, he advised members how to calculate their post-war subscriptions to the organisation:

> Think of a number. This might or might not be your pre-war subscription. Now add 10/9 or 21/6, according to whether you are over the age limit or not, multiply by 53 and divide by 47, on account of reduced working hours, and add 12½% war bonus. It is always to be understood that this division by 47 is optional and when it is not made, you still have the privilege of taking away the number you thought of first.[15]

Henry married Margaret Moore in 1903, the same year that he formed his own company, and the couple had two sons. Like Henry's sister, Kate, Margaret also trained as a kindergarten teacher at the Froebel Educational Institute and was active after her marriage as a committee member of the Michaelis Guild (an organisation for former students). Henry continued to enjoy sailing, and in later life was a keen horseman. At home, he ate his toast and marmalade with a knife and fork – and ruled the roost. Some years after Margaret's death, he remarried. He died in 1963. An obituary appeared a few months later, describing his personal qualities: 'No one who met him', it ran, 'could be unaware that he was a man of principle and personality.'[16]

The Cash family seem to have had little to do with Harold after 1907. Harold's second marriage proved quarrelsome and though the family may have believed it would provide good 'cover' for his irregular situation, it did not bring him peace. The first child, Harold Siegfried (known to the family as Sieg) was born with physical and intellectual disabilities, but unlike most of their contemporaries the couple did not place him in an institution. Instead, they cared for

Illustration 15.2   Henry Cash on 'Peggy', in about 1948. The photograph was taken at Wooton-Courtney, Exmoor, Somerset

him at home and kept him within the family. Harold's daughter, Helen, later recalled that her parents loved each other but her father's incessant drinking brought strain and distress, especially for her mother: 'Mummy was always the same loving mother but daddy was two people – an affectionate witty father and a stranger who came home drunk at times.' The family waited 'on tenterhooks on a Friday evening to see whether he came home with the house-keeping money or whether he came home squiffy having drunk most of the wages. It was especially bad at Christmas when he was likely to have celebrated more, while the home needs were greater.'[17]

Eventually, Harold lost his job and the money ran out. Over the next few years, the children were sent to live with various friends and relatives and the

Illustration 15.3 Harold with his youngest child, David, who was born in 1917

family was not reunited until Dorothy had successfully completed her train-ing as a midwife.[18] By now, they had moved back to Croydon. Harold was employed again, albeit on a lower salary, and with Dorothy's earnings too, they could settle at last. Somehow, Harold managed to curb his drinking. As grandchildren came along, he formed affectionate bonds with them. He is remembered now as a calm, benevolent presence in the family midst, sitting in a favourite chair, a kind of captain's chair, always reading and wearing a beret (to hide his baldness). When the children played hunt the thimble, he sat in the middle of the room to keep out of the way, and would offer his hat as a hiding place. He liked to sing early in the morning when he came downstairs to make Dorothy a cup of tea or when he was in the house on his own. In his later years he was lame and shuffled from room to room with the aid of a stick. He lived the rest of his life with the ignominy of not having been able to support his own family but seemed to grow stoical. He remained a committed socialist and he never spoke of his feelings for Grace. He died in 1952.

Walter, like Harold, lived into his eighties. After Joan died, he continued as the local doctor in Havelock North. Renée had left school and now, in her mother's absence, she helped in the house and was Walter's reluctant hostess when guests came. Walter handed his practice over to Tony in 1936, and then a few years later, after the outbreak of the Second World War, he returned to medical duties, this time at a military convalescent hospital in Rotorua. By 1949, now in his early seventies, he had fully retired from medicine and was back in Gisborne, living on the fringes of the town. What memories did he have of these streets, the landmarks and people in the place where he and Joan had lived some thirty years before? And what had brought him back?

The unlikely story concerns an Anglican vicar. It appears that Walter employed a housekeeper in Havelock North, sometime after Joan became ill, and that this woman, Kitty Hall, was the wife of the Reverend Alfred Hall, a local churchman. To Renée's acute discomfort, her father drew close to Kitty, whose husband was away in Tauranga. Kitty's marriage was not what it may have seemed. Though the couple had two sons, Walter's family believe the vicar was gay and his marriage little more than a respectable cover story.

Walter was willing to ignore social convention if the situation required it. With his gracious, courtly bearing, he was attractive to women and, after all, it was not the first time he had rescued someone else's wife. After the Second World War ended, he had left Rotorua to join Kitty in Gisborne, where her husband had taken a new post. The three of them now lived together, in a beautiful, colonial-style house known as Te Rau Kahikatea. Built in the 1870s, the house was a palatial vicarage – and with its bargeboards, gables and ornate verandahs, it remains a striking example of Gothic revival architecture. Four or five bedrooms, and several reception and service rooms were spread over two floors – and here Walter lived with Kitty, and the man his family came to call 'Canon Bob', until the 1950s.

Kitty was well-liked by Walter's family. She was almost twenty years his junior, tall and statuesque, kind and loving. Several times, the family visited Walter in Gisborne and he made the trip to England too, to see Colin who was now married and living there. On these occasions Walter once again enjoyed some sailing but he made no contact with the Cash family, or with former friends like Stanley Carr.

Then, in 1954, the Reverend Hall was sent to continue his ministry in Paeroa, a small windswept town located some two hundred miles north of Gisborne, at the foot of the Coromandel Peninsula. Once again, Walter and Kitty went with him and the three took up residence in the vicarage. How did the vicar and his wife explain Walter's presence to curious locals? Was he introduced as a family friend or lodger, perhaps? As it happened, the Reverend Hall's ministry in Paeroa was short and he died only two years after his arrival. In the same year, Walter and Kitty between them bought a modest, wooden bungalow that still stands in the town. They lived there until Walter's death the following year, on 21 August 1957, at the age of 81. They did not marry. Walter was buried in Paeroa, in an unmarked grave on a steep hillside in the local Pukerimu Cemetery, in a section reserved for those with no religious affiliation.

In 1959, just two years after his father's death, Colin, the younger of the twins, began digging into the family's past. All three children had by now been told the truth. Tony had been first to learn in 1934 that his parents were not legally married. That was the year he completed his medical training in Edinburgh and announced his engagement to Noel Lawson, a young woman he had known in New Zealand who was now living in Britain. When Walter heard the news, he wrote to Tony's prospective father-in-law and told him Joan's story, in case he should feel the match was inappropriate. Tony was stunned – but the marriage went ahead as planned. Colin took a more light-hearted view of their parents' escapade, and for many years enjoyed dining out on the story. In England he obtained details of the staged drowning from Dr Francis Carr and from a now elderly Charles Mitchiner. He also contacted Henry Cash, by now well into his eighties, and received the brush-off. Grace's young brother, said Colin, 'became rather crusty'.[19] He made it clear he did not wish to meet his sister's children or enter into correspondence with them. 'As far as this family is concerned, our sister is dead and has been for some time', he said.

Colin afterwards wrote to Renée in Christchurch, where she now lived with her husband, an academic. He told his sister he believed that Henry Cash had a 'guilty conscience about the way he behaved in the whole incident'.[20] Yet no details about Henry's involvement have emerged. Described as 'a man of principle' by his colleagues[21] there is no doubt he kept his word and told his sister's secret to no one, not even his own children. But how did this principled man regard Grace's actions? Did he feel responsible for having brought Grace and Walter together in 1899 perhaps, for having left them alone on the boat more than once? Did he disapprove of their decision or did he support them? Was it Henry perhaps who sailed into the bay at Arzon in 1907 and picked his sister up out of the water? His role in her disappearance, if indeed he had one, remains a mystery.

Back in Havelock North, Tony struggled for some years to come to terms with his parents' actions. It was not that he did not love both of them, as he explained to Renée, or that he disapproved of what they had done. But as the local doctor and a public figure in a small village, he was 'too close to it' and would have found the 'sneering remarks' about two people he admired and respected quite intolerable. It was easier, he said, for Colin and Renée because they had moved away. Tony tended to bottle things up and he had a less progressive outlook than either Renée or Colin.[22] Political differences had led to the brothers becoming estranged in later life. But Renée understood Tony's feelings. Now in her eighties, she reminded her brother of their happy childhood and their happy family life. 'We were exceptionally blessed in our parents', she said.[23]

Notes

Prologue

1. *L'Avenir du Morbihan*, 28 August 1907.
2. *The Times*, 6 September 1907, p. 4, col. (e).
3. 'ILP Notes', *The Croydon Times*, 14 September 1907, p. 8, col. (e).
4. *The Englishwoman's Review*, 15 October 1907, pp. 267–8.
5. Women's Industrial Council, *Thirteenth Annual Report 1906–07* (London: Women's Industrial Council), p. 34.

1 Radicals in Suburbia

1. The term *fin de siècle* refers to the end of the nineteenth century. Some scholars of the period focus on the 1890s; others define it more loosely and encompass the 1880s and the early part of the twentieth century as well.
2. The advent of the 'New Woman' is discussed more fully in Chapter 4, 'Another Word for Suicide'.
3. The specific circumstances and concerns of the lower middle classes in Britain at this time are discussed in G. Crossick (1977) (ed.), *The Lower Middle Class in Britain 1870–1914* (London: Croom Helm).
4. The Girls' Public Day School Company (GPDSC) was set up in 1872 and still exists as 'The Girls' Day School Trust'. For more details, see Chapter 2, 'Learning Curves'.
5. J. Kamm (1971), *Indicative Past: A Hundred Years of the Girls Public Day School Trust* (London: George Allen & Unwin), p. 70.
6. The house was compulsorily purchased and demolished in 1930.
7. A. C. Doyle (1893), *The Great Shadow* and *Beyond the City* (Bristol: J. W. Arrowsmith), pp. 162–3.
8. Decimus Burton Road is now part of Grange Road. Decimus Burton was a protégé of John Nash (best remembered for his terraces in London's Regent's Park).
9. H. Corke (1975), *In Our Infancy: An Autobiography, Part 1: 1882–1912* (Cambridge: Cambridge University Press), p. 97.
10. D. H. Lawrence came to teach in Croydon in 1908.
11. The figure for James' income is a speculation based on data reported in M. Heller (2008), 'Work, Income and Stability: The Late Victorian and Edwardian London Male Clerk Revisited', *Business History*, Vol. 50(3): 253–71.
12. K. Chorley (1950), *Manchester Made Them* (London: Faber & Faber), p. 149.
13. The National Archives (TNA), Board of Trade records, BT31/5294/36171 and BT31/31999/98494.
14. In 1918 another stationery company, John Dickinson & Co. Ltd, purchased the share capital of Millington's. The two companies continued to trade independently until 1932 when they finally amalgamated.
15. G. Grossmith and W. Grossmith [1892] (1984), *The Diary of a Nobody* (London: Elm Tree Books), p. 35.
16. Ibid., p. 37.

17. G. Gissing (1901), 'A Daughter of the Lodge', in A. Richardson (2002) (ed.), *Women Who Did: Stories by Men and Women, 1890–1914* (London: Penguin Books), pp. 270–1.
18. 'Chapel' was the term used to describe church buildings belonging to independent or Nonconformist religious groups.
19. Nonconformists were members of Protestant churches in England that dissented from the established Anglican Church.
20. Thornton Heath Congregational Church Archives (CLSL), PR2/7/3/1/1. Congregationalism is a protestant movement that stresses the independence and autonomy of each individual church and the right of each congregation to decide their own affairs without having to submit to a higher authority (such as a bishop or synod).
21. These examples are taken from a programme for a fund-raising bazaar held in 1902. CLSL, PR2/7/1/7/5/1.
22. The beliefs and activities of 'The Fellowship of the New Life' are explained fully in Chapter 5, 'Fellowship is Heaven'.
23. Eventually, William Jupp became the leader of a free religious movement in Croydon.
24. P. Bailey (1999), 'White Collars, Gray Lives? The Lower Middle Class Revisited', *Journal of British Studies*, Vol. 38(3): 273–90 (276).

2   Learning Curves

1. The Education Act of 1870 had made a state system of elementary education (i.e. initial education for working-class pupils) possible but attendance was not made compulsory until 1880 and elementary schools were not free until 1891. A few working-class children were able to go to secondary schools with the help of scholarships, but generally post-elementary education was the preserve of the middle classes. All secondary schools were fee-paying and would remain so until 1944. A secondary school place for all children was not compulsory until after 1944.
2. M. Arnold (1868), *Schools and Universities on the Continent* (London: Macmillan), p. 296.
3. In 1890 the institution was renamed 'Westminster City School' and this is the name it has today. The four original schools were Emanuel Hospital, St Margaret's Hospital, Palmer's School and Emery Hill's School.
4. City of Westminster Archive Centre (CWAC), Records of Westminster City School, ACC1657.
5. R. Carrington (1983), *Westminster City School and its Origins* (London: United Westminster Schools Foundation), p. 27.
6. Further details of Henry's professional career are given in Chapter 15, 'After Lives'.
7. Details about Harold's school activities from 1884 to 1887 are taken from the school magazine, *The Novocastrian*, which was consulted at the Royal Grammar School, Newcastle.
8. Details of the Oakeshott family's involvement with the university extension movement are taken from the local newspaper for the 1880s and 1890s, the *Sunderland Daily Echo and Shipping Gazette*.
9. O. Browning (1887), 'The University Extension Movement at Cambridge', *Science* (21 January): 61–3 (63).
10. Ibid., p. 61.
11. The Fabian Society is a British socialist organisation, founded in 1884. Both Joseph Francis Oakeshott and his younger brother, Harold, were members of it in the 1890s.

More details about their involvements in socialist groups are given in Chapter 5, 'Fellowship is Heaven'.

12. 'Fabian Picnic', *Sunderland Daily Echo and Shipping Gazette*, 5 August 1891, p. 4.
13. *Sunderland Daily Echo and Shipping Gazette*, 8 May 1891, p. 3.
14. *Newcastle Courant*, 15 December 1888, p. 4.
15. *Sunderland Daily Echo and Shipping Gazette*, 11 July 1893, p. 3.
16. F. Cobbe (1904), *Life of Frances Power Cobbe As Told by Herself* (London: Swan Sonnenschein & Co.), p. 58.
17. The motto was quoted in a tribute to Dorinda Neligan (Grace's headmistress) in 1914. See 'In Loving Remembrance of Miss Neligan', *The Croydon Advertiser and Surrey County Reporter*, 25 July 1914, p. 7.
18. J. Sondheimer and P. Bodington (1972) (eds), *GPDST 1872–1972: A Centenary Review* (London: GPDST), p. 10.
19. Ibid., p. 45.
20. Croydon High School Archives, CHS.
21. Letter to Maria Grey, one of the founders of the GPDSC, quoted in J. Kamm (1971), *Indicative Past: A Hundred Years of the Girls' Public Day School Trust* (London: George Allen & Unwin), p. 61.
22. Biographical details are from 'Dorinda Neligan 1833–1914', undated typescript biography by Eleanor Roper. Institute of Education Archives (IOE), GDS/12/7/1.
23. M. V. Hughes (1946), *A London Girl of the 1880s* (Oxford, New York, Toronto and Melbourne: Oxford University Press), p. 122.
24. On one occasion, when Neligan refused to pay her rates, her silver teapot was 'distrained'. In 1910 she was arrested for supposedly assaulting a policeman but she was never charged or imprisoned.
25. *The Croydon Advertiser and Surrey County Reporter*, 25 July 1914, p. 7.
26. 'A Short Account of the First Twenty Five years of Croydon High School' (n.d.), edited by senior pupils, unpublished, IOE, GDS/23/2/6/2.
27. See E. Jordan (1991), '"Making Good Wives and Mothers"? The Transformation of Middle-Class Girls' Education in Nineteenth-Century Britain', *History of Education Quarterly*, Vol. 31(4): 435–62.
28. *The New Guide to Ramsgate, Margate, Broadstairs and St. Peter's* (1886) (Ramsgate: S. R. Wilson), p. 69.
29. J. Perkin (1993), *Victorian Women* (London: John Murray), p. 34.
30. D. Copelman (1986), '"A New Comradeship between Men and Women": Family, Marriage and London's Women Teachers, 1870–1914', in J. Lewis (ed.), *Labour and Love: Women's Experience of Home and Family 1850–1940* (Oxford: Blackwell).

3 Finding Their Own Way

1. W. Robinson (2003), *Pupil Teachers and their Professional Training in Pupil-Teacher Centres in England and Wales, 1870–1914* (Lewiston, NY: The Edwin Mellen Press), p. 155.
2. Report from the School Board for London, year ending 25 March 1888. London Metropolitan Archives (LMA), SBL/1484, p. 65.
3. H. Dunn (1869), *Teaching: Its Pleasures, its Trials... ... Being Counsels for Pupil-Teachers, Governesses ...* (London: Simpkin Marshall & Co.), p. 111.
4. For Jessie's testimonial from Warple Way School, see the Stockwell College records in the British and Foreign School Society Collection, held at Brunel University,

BFSS/STC/Apps/1889/Jessie Cash. For relevant student lists and Jessie's examination results, see BFSS/1/1/6 [AR 1892].

5. D. H. Lawrence [1915] (1995), *The Rainbow* (London: Wordsworth Editions), p. 318.
6. H. Corke (1975), *In Our Infancy* (Cambridge: Cambridge University Press), p. 103.
7. Ibid., p. 117.
8. Ibid., p. 132.
9. Ibid., p. 141.
10. Sydenham Road Girls' Schools Archives, Croydon Local Studies Library (CLSL). For the Girls' School log book (3 April 1905–8 March 1917) see SCH135/1/1/4; for relevant administrative papers for the schools, see CBC/3/2/85 and CBC/3/2/86.
11. Ibid., SCH135/1/1/4.
12. Jessie's salary is equivalent to just over £12,000 today in terms of its purchasing power and it was less than half the salary that her younger sister, Grace, received in the same year on her appointment to the London County Council (see Chapter 9, 'Girls in Trades').
13. Board of Trade records, TNA, BT31/2641/14019.
14. See P. Weston (2002), *The Froebel Educational Institute: The Origins and History of the College* (Roehampton: University of Surrey).
15. The Froebel Educational Institute moved to Roehampton, Surrey, in 1921. Kate Cash was eventually employed there as a teacher (for more details of her academic career, see Chapter 15: 'After Lives').
16. Froebel Educational Institute records, The Froebel Archive for Childhood Studies, University of Roehampton (UR), NFF/1.
17. Quoted in an obituary in *The Link*, 1950, No. 40, p. 24. This newsletter for FEI alumni was consulted in The Froebel Archive for Childhood Studies (UR).
18. Quoted as part of an anonymous tribute entitled 'Madame Michaelis', *Child Life*, Vol. 7(26) (15 April 1905), p. 64.
19. A. Phillips (1979) (ed.), *A Newnham Anthology* (Cambridge: Cambridge University Press), p. 36.
20. Dorinda Neligan was anxious not to take too many boys into the kindergarten class and to commit resources to teaching them because when they reached school age, they would go elsewhere. She believed the kindergarten class should 'feed' the main school.
21. 'Cambridge Letter' (1892) (Newnham College Club, private circulation), pp. 5–6.
22. London University granted degrees to women from 1878 but Cambridge held out until 1948, and even then retained the power to limit their numbers.
23. Phillips, *Anthology*, p. 32.
24. Gillian Sutherland, 'Clough, Anne Jemima (1820–1892)', *Oxford Dictionary of National Biography* (Oxford University Press, 2004), p. 3. Available at: http://www.oxforddnb.com/view/article/5710 (accessed 2 March 2011).
25. Quoted in C. Dyhouse (1981), *Girls Growing Up in Late Victorian and Edwardian England* (London: Routledge & Kegan Paul), p. 77.
26. Phillips, *Anthology*, p. 38.
27. Ibid., p. 46.
28. Ibid., p. 57.
29. Ibid., p. 69.
30. Confirmation that Grace's daughter was named after her close friend at Cambridge is contained in unpublished family correspondence.
31. CHS, administrative records.
32. Details of Jessie's later teaching career are given in Chapter 15, 'After Lives'.

4 'Another Word for Suicide'

1. In the south-east of England in 1894, only 14 marriages out of every 100 were conducted in civil registries. See O. Anderson (1975), 'The Incidence of Civil Marriage in Victorian England and Wales', *Past & Present*, No. 69, p. 55.
2. Noted in B. Brookes (1986), 'Women and Reproduction, 1860–1939', in J. Lewis (ed.), *Labour and Love: Women's Experience of Home and Family 1850–1940* (Oxford: Basil Blackwell).
3. C. Seymour-Jones (1992), *Beatrice Webb: Woman of Conflict* (London: Pandora), p. xii.
4. *Beatrice Webb Diary*, Volume 14, entry for 1 December 1890. The diaries are available online at http://www.digital.library.lse.ac.uk.
5. A full account of Grace Oakeshott's work for the Women's Industrial Council is given in Chapter 8, 'Behind Closed Doors'.
6. Noted in Seymour-Jones, *Beatrice Webb*, p. 188.
7. The activities and beliefs of this organisation (as well as Harold's contribution to it) are described fully in Chapter 5, 'Fellowship is Heaven'.
8. Quoted in G. Frost (2008), *Living in Sin: Cohabitating as Husband and Wife in Nineteenth-Century England* (Manchester and New York: Manchester University Press), p. 205.
9. S. Bowen [1941] (1984), *Drawn from Life* (London: Virago), p. 54.
10. E. Hobsbawm [1987] (1994), *The Age of Empire, 1875–1914* (London: Abacus), p. 195.
11. These meetings were reported in *Seed-time*, the journal of the Fellowship of the New Life, London School of Economics (LSE), FABIAN SOCIETY/E/117. See *Seed-time*, No. 28 (April 1896), pp. 14–15 for a summary of Clementina Black's talk on the work of the Women's Industrial Council; *Seed-time*, No. 25 (July 1895), p. 15 for a summary of a talk by Amie Hicks on 'Women and the Trade Unions'; *Seed-time*, No. 6 (October 1890), pp. 10–11 for an article by Clementina Black on 'The Ethics of Shopping'.
12. M. Caird (2000), 'Marriage (1888)', Extract in S. Ledger and R. Luckhurst, *The Fin de Siècle: A Reader in Cultural History c.1880–1900* (Oxford: Oxford University Press), p. 79.
13. Sarah Grand was best known for her controversial novel *The Heavenly Twins* (1893) in which her three heroines struggle in different ways with the constraints of Victorian womanhood.
14. C. Perkins Gilman (2002), 'The Yellow Wallpaper (1892)', in A. Richardson (ed.), *Women Who Did: Stories by Men and Women, 1890–1914* (London, Penguin Books), p. 32.
15. Harold's father, Joseph, had died in 1893 leaving his estate (gross value £514 8s. 0d., equivalent today to £49,780 in terms of its purchasing power) to his wife Eliza. It seems likely that Harold was able to use some of the inheritance to build Downside Cottage, perhaps on the understanding that at least initially his mother and his sister, Mary, would live there too.
16. H. Oakeshott (n.d.), 'The Ethics of Marriage', unpublished family papers.
17. H. Oakeshott (1897), 'The Fellowship Gild', *Seed-time*, No. 31 (January), pp. 8–13.
18. Brookes, 'Women and Reproduction', p. 152.
19. Hobsbawm, *The Age of Empire*, p. 192ff.
20. See Brookes, 'Women and Reproduction'.
21. A. James Hammerton (1990), 'Victorian Marriage and the Law of Matrimonial Cruelty', *Victorian Studies* Vol. 33(2): 269–92 (270).
22. Seymour-Jones, *Beatrice Webb*, p. 236.
23. It was not until 1923 that Parliament permitted a wife to present a divorce petition on grounds of adultery alone and it was 1969 before the Divorce Reform Act allowed

couples to divorce if they had been separated for two years, simply on grounds of the irretrievable breakdown of the marriage. Men were still favoured in the distribution of assets, however, which were not split more fairly until the 1990s.

24. W. Morris [1912] (2005), *The Collected Works of William Morris*, Vol. XVI (London: Elibron Classics), pp. 56–7.
25. For further background to the Fabian dispute, see M. Drabble, 'Introduction', in H. G. Wells [1909] (2005), *Ann Veronica* (London, Penguin Books), pp. xiii–xxxii.

5 'Fellowship is Heaven'

1. Morris, *The Collected Works*, p. 4.
2. Ibid., p. 14.
3. Ibid., p. 211.
4. J. Oakeshott (1897), 'William Morris', *Seed-time*, No. 31, p. 1.
5. Accounts of these times are strewn with the names of those who were to become prominent in the socialist movement of the late 19th and early 20th centuries (Beatrice and Sidney Webb, H. G. Wells, Havelock Ellis, Edward Carpenter, J. Ramsay MacDonald). This story concerns some of the minor stars, less written about but no less hardworking or committed than their better-known comrades.
6. Joseph Francis Oakeshott was a member of the Fabian Executive from 1890 until 1902. From 1880 he was employed by the Inland Revenue, at Somerset House in London. With an expertise in statistics he was particularly influential in the Fabian Society's attempts to apply socialism to taxation. Joseph wrote four Fabian tracts, but only one ('The Humanizing of the Poor Law', Fabian Tract No. 54) was issued in his name. His second son Michael (b.1901) had a distinguished career as a political philosopher. LSE, FABIAN SOCIETY.
7. William Jupp was a close friend of Adams. Later he founded a free religious movement in Croydon and became prominent in the Croydon Brotherhood Church.
8. W. J. Jupp (1919), *Worlds not Realised: A Study of the Human Personality in the Light – or Shadow – of its Unfulfilled Possibilities* (London: Headley Bros.), pp. 11–12.
9. The journal was originally called *The Sower*. After its first issue, it was renamed *Seed-time*. Maurice Adams was its editor from its inception in 1889 until the Fellowship closed in 1898.
10. M. Adams (1889), 'Editorial Preface', *The Sower*, No. 1, pp. 1–2.
11. Ethical socialism is a political philosophy that advocates socialism on moral and ethical rather than economic or materialist grounds.
12. W. J. Jupp (1918), *Wayfarings: A Record of Adventure and Liberation in the Life of the Spirit* (London: Headley Bros.), p. 84.
13. J. Oakeshott (1896), 'The New Fellowship: A Retrospect', *Seed-time*, No. 29, pp. 1–2.
14. No details about James' talk appear in the Fellowship's records.
15. The Croydon Brotherhood Church opened in 1894 and its first pastor was John Kenworthy, a follower of Tolstoy's teachings. It was modelled on the first Brotherhood Church, established in Hackney in London in 1891, by John Bruce Wallace. In Croydon there was an initial focus on small-scale co-operative industries and the group produced a newspaper called *The New Order*. The Croydon Brotherhood Church became increasingly and explicitly Tolstoyan during the 1890s.
16. N. Shaw (1935), *Whiteway: A Colony on the Cotswolds* (London: C. W. Daniel & Co.), p. 21.
17. H. Oakeshott (1897), 'The Fellowship Gild', *Seed-time*, No. 31, p. 9.

18. Ibid., p. 10.
19. Ibid., p. 11.
20. Ibid.
21. H. Oakeshott (1898), 'The Worth of the New Fellowship', *Seed-time*, No. 34, p. 13.
22. Ibid., p. 14.
23. Ibid., p. 13.
24. Ibid.
25. M. Adams (1898), 'Editorial Note', *Seed-time*, No. 34, p. 1.
26. Oakeshott, 'The Worth', p. 14.
27. Lees was an associate of Eleanor Marx and, like her, an early member of the Fabian Society. According to a biography of Havelock Ellis, to him Lees appeared vivacious, competent and efficient but a troubled childhood had left her with an underlying vulnerability. See P. Grosskurth (1980), *Havelock Ellis: A Biography* (London: Allen Lane).
28. E. Lees (1890), 'Woman and the New Life', *Seed-time*, No. 3, p. 5.
29. E. Havelock Ellis (1909), *Attainment* (London: Alston Rivers).
30. The principles of communism were formally abandoned at Whiteway in 1902 but a residential community still exists there with more than 150 colonists of all ages, and over 60 homes. See http://diggersanddreamers.org.uk/index.php?fld=initial&val=W&one=dat&two=det&sel=whiteway (accessed 20 January 2015).
31. Shaw, *Whiteway*, p. 56.
32. Corke, *In Our Infancy*.
33. See H. Mason [née Oakeshott] (2005), *What Time is it now? My Life as a Teacher* (Porthleven: Kirrier-Porthleven), the autobiography of Harold's daughter from his second marriage, who grew up in Downside Cottage.
34. All the diary extracts and the map in Illustration 5.1 are taken from unpublished family sailing logs.
35. The logbooks provide a first-hand account of a number of sailing trips led by Henry Cash in the late nineteenth and early twentieth centuries, and they form a kind of collective diary of experiences on board. Although authorship is not always directly attributable to one individual, often it appears that the person acting as 'skipper' or captain for that trip has completed the entries. Handwriting comparisons have confirmed that for the trips mentioned here this was usually Henry, and that when he was absent for part of the East Coast trip in 1900, Walter took over the logbook, leading to distinct changes in recording style and tone.
36. All those present on any cruise were given nautical titles ('skipper', 'able-bodied seaman', 'first mate' and so on) and the log books often refer to them in this way. At other times, just their initials were used.
37. Also spelt 'Elberry'.
38. The bathing house first appeared on maps of the area in the 1860s. At various times in its past the roof was thatched.
39. F. Cowper (1928), *Yachting and Cruising for Amateurs* (London: The Bazaar, Exchange & Mart), p. 103.
40. G. Davies (1886), *Practical Boat Sailing for Amateurs* (London: L. Upcott Gill), p. 1.

6   Answering the Call

1. For an official history of the CMS, see E. Stock (1899), *History of the Church Missionary Society: Its Environment, Its Men, and Its Work*, Volumes 1–4 (London: Church Missionary Society).
2. C. Bronte [1847] (1994), *Jane Eyre* (London: Penguin Books), p. 35.

3. M. Rutherdale (2002), *Women and the White Man's God: Gender and Race in the Canadian Mission Field* (Vancouver: UBC Press), p. 9.
4. F. Du Vernet (1911), 'News from the Front: Diocese of Caledonia', *Across the Rockies* (October). Quoted in Rutherdale, *Women and the White Man's God*, p. 10.
5. Most biographical details are from C. Mockridge (1896), *The Bishops of the Church of England in Canada and Newfoundland* (Toronto: F. N. W. Brown), which was published while Reeve was still active in the missionary field.
6. W. Bompas (1888), *Diocese of Mackenzie River* (London: SPCK).
7. W. Reeve (1869), *Fort Simpson Journal*, 30 August, p. 26. Cadbury Research Library, University of Birmingham (UBCRL), CMS/B/OMS/C C1 0 54/10.
8. For a summary history of the Hudson's Bay Company, see Hudson's Bay Company Archives at http://www.gov.mb.ca/chc/archives/hbca/about/hbc_history.html (accessed 20 January 2015).
9. Fort Simpson was the headquarters of the Anglican mission and the earliest station to be established in the Mackenzie River District.
10. In standard ethnological classifications, Canada's First Nations of Dene (the Indians of the Northwest Territories) are grouped as Chipewyans, Slaveys, Dogribs, Mountain Indians, Bearlake Indians, Hares and Gwich'in. The other native peoples of the Northwest Territories are the Northern Metis (of mixed European and First Nations heritage) and the Inuit. See J. Helm (2000), *The People of the Denendeh: Ethnohistory of the Indians of Canada's Northwest Territories* (Montreal and Kingston, London, Ithaca: McGill-Queen's University Press).
11. Reeve (1869), *Journal*, 2 September, p. 2. UBCRL, CMS/B/OMS/C C1 0 54/10.
12. S. Archer (1929), *A Heroine of the North: Memoirs of Charlotte Selina Bompas (1830–1917)* (London: SPCK), p. 24.
13. Reeve (1869), *Journal*, 24 December, p. 19. UBCRL, CMS/B/OMS/C C1 0 54/10.
14. Ibid., p. 15.
15. W. Reeve (1874), Letter to CMS, 22 June, p. 2. UBCRL, CMS/B/OMS/C C1 0 54/4.
16. Reeve (1878), *Journal*, 9 June. UBCRL, CMS/B/OMS/C C1 0 54/16.
17. Archer, *A Heroine of the North*, p. 23.
18. The Athabasca diocese was later divided, and William Bompas then became the first Bishop of Mackenzie River. In 1891, when Bompas moved to the Yukon, William Day Reeve was ordained as a bishop and took over the Mackenzie River post.
19. Bompas' life of devotion, privation and physical endurance would eventually be celebrated in Victorian times as a model for all Christians. For a biography, see K. Abel (2000), *Bompas, William Carpenter*, Dictionary of Canadian Biography Online, available at http://www.biographi.ca/en/bio/bompas_williaarpentem_cr_13E.html (accessed 20 January 2015).
20. Archer, *A Heroine of the North*, p. 26.
21. Ibid., p. 25.
22. Ibid., pp. 26–7.
23. Reeve (1870), *Journal*, 30 April, p. 33. UBCRL, CMS/B/OMS/C C1 0 54/10.
24. Emily's correspondence is summarised in B. Kelcey (2001), *Alone in Silence: European Women in the Canadian North before 1940* (Montreal and Kingston, London, Ithaca: McGill-Queen's University Press), p. 39.
25. Archer, *A Heroine of the North*, p. 32.
26. Ibid.
27. F. Russell (1898), *Explorations in the Far North: Being the Report of an Expedition under the Auspices of the University of Iowa during the years 1892, '93 and '94* (Iowa City: University of Iowa), p. 69.

28. Reeve (1875), Letter to CMS, 19 June. UBCRL, CMS/B/OMS/C C1/054/6.
29. W. Reeve (1876), *Annual Letter*, 27 June. UBCRL, CMS/B/OMS/C C1 0 54/22.
30. Kelcey, *Alone in Silence*. See Chapter 7 for an account of birthing practices in the Canadian Arctic at this time.
31. See K. Abel (2005), *Drum Songs: Glimpses of Dene History* (2nd edn.) (Montreal and Kingston, London, Ithaca: McGill-Queen's University Press).
32. W. Reeve (1877), letter to his sister, dated April (unpublished family papers).
33. Archer, *A Heroine of the North*, p. 48.
34. Reeve (1877), *Annual Letter*, 2 July. UBCRL, CMS/B/OMS/C C1 0 54/23.
35. Reeve (1876), *Annual Letter*, 27 June, p. 3. UBCRL, CMS/B/OMS/C C1 0 54/22.
36. Abel, *Drum Songs*, Chapter 6.
37. Reeve, *Annual Letter*, 2 July. UBCRL, CMS/B/OMS/C C1 0 54/23.
38. Ibid.
39. Reeve (1870), *Journal*, 5 June, p. 48. UBCRL, CMS/B/OMS/C C1 0 54/11.
40. K. Temprano (2011), *Rheumatoid Arthritis and Pregnancy*, http://www.emedicine.medscape.com (accessed 18 August 2011).
41. Reeve (1870), *Journal*, 1 January, pp. 20–1. UBCRL, CMS/B/OMS/C C1 0 54/10.
42. *The Church Missionary Gleaner*, Vol. 33(389) (May 1906): 67. The magazine was consulted at UBCRL.
43. Isaac Stringer had a long career in the Western Arctic. After an arduous journey in 1909, when Stringer and his companion were lost in the mountains for 51 days, and were forced to eat shoe leather, he became known as the 'Bishop who Ate his Boots'. For details, see http://www.museevirtuel-virtualmuseum.ca/sgc-cms/expositions-exhibitions/eveque-bishop/english/fullstory.html (accessed 20 January 2015).
44. Reeve (1893), *Annual Letter*, 30 November. Minnesota Historical Society (MHS), P2137.

7   Not Much 'Home' About It

1. F. Shaw [1878] (1966), *Castle Blair: A Story of Youthful Days* (London: Rupert Hart-Davis), p. 33.
2. *St. Michael's Old Members Magazine*, March 1928, No. 28, UBCRL, CMS/ACC 922, p. 398.
3. M. Cosh (2005), *A History of Islington* (London: Historical Publications), p. 169.
4. B. Richardson (2009) (ed.), *Sons of Missionaries: Recollections by Boarders at Eltham College* (Eltham: School for the Sons of Missionaries), pp. 6–7.
5. Admission form for CM Home. UBCRL, CMS/G/AMc 11/4, p. 2.
6. See Rules for the CM Children's Home, including outfit lists. UBCRL, CMS/G/AMc 10.
7. *St. Michael's Old Members Magazine*, p. 379.
8. C. Witting (1952) (ed.), 'The Glory of the Sons: A History of Eltham College, School for the Sons of Missionaries', Volume 1, *Eltham College Past and Present* (London: Board of Governors, Eltham College), p. 164.
9. Church Missionary Children's Home, Regulations 1888. UBCRL, CMS/G/AMc 10.
10. *St. Michael's Old Members Magazine*, p. 380.
11. Church Missionary Children's Home, Regulations 1888. UBCRL, CMS/G/AMc 10, p. 2.
12. Church Missionary Children's Home, Visiting Regulations 1875. UBCRL, CMS/G/AMc 10.
13. *St. Michael's Old Members Magazine*, p. 399.
14. Witting, 'The Glory of the Sons', p. 163.

15. Quoted in V. Brendon (2006), *Children of the Raj* (London: Phoenix), p. 205.
16. *St. Michael's Old Members Magazine*, p. 379.
17. Ibid., p. 380.
18. In all likelihood, this church was St Mary's in Upper Street, Islington, which was a key focus for evangelicalism throughout the nineteenth century. The church was bombed in 1940 and, apart from the tower, was totally destroyed. It was rebuilt in 1956, in a design that incorporated the original tower.
19. *St. Michael's Old Members Magazine*, p. 380.
20. Ibid., p. 398.
21. Church Missionary Children's Home Report, July 1882. UBCRL, CMS/G/AMc 3, p. 3.
22. D. Coleman and J. Salt (1992), *The British Population: Patterns, Trends and Processes* (Oxford: Oxford University Press), p. 52.
23. Church Missionary Children's Home Report, 1886. UBCRL, CMS/G/AMc 3, p. 6.
24. *St. Michael's Old Members Magazine*, p. 329.
25. The present day St Michael's in Limpsfield, Surrey, is a Grade II listed building and now comprises privately-owned apartments.
26. E. Wright (1937), *The Children of Thy Servants: Memories of Highbury and Limpsfield, 1850–1937* (Surrey: St Michael's, Limpsfield), p. 44.
27. *St. Michael's Old Members Magazine*, p. 328.
28. Ibid.
29. Ibid.
30. Ibid.
31. M. Wright (1996), *Havelock North: The History of a Village* (Hastings: Hastings District Council), p. 148.
32. For a history of the school, see A. Lace (1968), *A Goodly Heritage: A History of Monkton Combe School, 1868–1967* (Bath: Monkton Combe School).
33. Historical records for Monkton Combe School (including the Prefects' Minutes) were consulted in the school archives (MCS).
34. As shown in Illustration 7.2, the rugby ball Walter's team used was more spherical than it is today. It only became ovoid when the emphasis of the game moved towards handling and away from dribbling.
35. Lace, *A Goodly Heritage*, p. 82.
36. Richardson, *Sons of Missionaries*, p. 8.

8   Behind Closed Doors

1. C. Black (1900), *The Rhyme of the Factory Acts* (London: Women's Industrial Council), p. 2.
2. As well as working for the Women's Industrial Council, Clementina Black was a prolific writer, a Fabian, and an active campaigner for women's suffrage.
3. For example, both Beatrice Webb (the social reformer and Fabian) and Margaret MacDonald (the wife of Ramsay McDonald) who were members of the WIC at this time had private incomes.
4. 'The "Speaker" Reports', *WIN*, December 1897, p. 7.
5. M. Bondfield (1949), *A Life's Work* (London: Hutchinson), p. 33.
6. *Seed-time*, No. 25, July 1895, p. 15.
7. *Seed-time*, No. 28, April 1896, p. 15.
8. 'Grace Marion Oakeshott', *WIN*, December 1907, p. 668.
9. 'Women's Trade Union Association', *The Times*, 27 November 1894, p. 12.

10. For a discussion of the largely middle-class appeal of the suffrage campaign in Britain at this time, see E. Hobsbawm (1987), *The Age of Empire, 1875–1914* (London: Abacus), Chapter 8, 'The New Woman'. Note, however, that some suffrage groups made efforts to attract working-class women and that there was variation by region. For example, in the North of England in the 1890s, there was a women's suffrage movement amongst mill hands and textile workers.

11. *WIC Annual Report*, 1900–1, p. 8.

12. See Charles Booth Online Archive, http://booth.lse.ac.uk/ (accessed 20 January 2015).

13. Quoted in J. Ramsay MacDonald (1929), *Margaret Ethel MacDonald* (London: Allen & Unwin) (6th edn.), p. 143.

14. G. Oakeshott (1903), 'Artificial Flower-Making: An Account of the Trade and a Plea for Municipal Training', *WIN*, No. 23, June 1903, p. 365.

15. Ibid., p. 366.

16. Ibid.

17. Ibid.

18. Note that Seebohm Rowntree (1871–1954), the second son of Joseph Rowntree, published a study into poverty in the early 1900s in which he calculated that in 1899 it cost 3*s*. 3*d*. a week to maintain an adult on a nutritionally balanced diet.

19. Oakeshott, 'Artificial Flower-Making', p. 367.

20. 'Home Work of Children', *WIN*, December 1897, p. 8.

21. 'A Tragic Ending', *WIN*, March 1898, p. 26.

22. For a detailed discussion of the workings of social class within the Women's Industrial Council, see G. Holloway (1995), *A Common Cause? Class Dynamics in the Women's Industrial Movement, 1888–1918* (unpublished PhD thesis, University of Sussex).

23. J. Hannam (2008), 'MacDonald, Margaret Ethel Gladstone (1870–1911)', *Oxford Dictionary of National Biography* (Oxford University Press, 2004), http://www.oxforddnb.com/view/article/45462 (accessed 16 July 2012).

24. Oakeshott, 'Artificial Flower-Making', p. 369.

25. Ibid., p. 372.

26. Ibid.

27. Grace's work in establishing the Trade Schools for Girls is the subject of Chapter 9, 'Girls in Trades'.

28. G. Oakeshott (1903), 'The Need for Social Investigation: Its Practical Bearing on the Work of the WIC', *WIN*, March, p. 360.

29. Ibid., pp. 361–2.

30. Ibid., p. 364.

31. G. Oakeshott (1908), *Reports on Women's Trades* (London County Council) [These reports were published after Grace's presumed death].

32. G. Oakeshott (1902), 'Women Polishers', *WIN*, March, pp. 285–92.

33. G. Oakeshott (1900), 'Women in the Cigar Trade in London', *The Economic Journal*, Vol. 10(40) (December): 572.

34. *WIC Annual Report*, 1904–5, p. 25.

35. L. Wyatt-Papworth (1914), 'The Women's Industrial Council: A Survey', *WIN*, No. 64, January, p. 208.

36. For a history of the work of the Women's Industrial Council (including some extracts from its reports), see E. Mappen (1985), *Helping Women at Work: The Women's Industrial Council 1889–1914* (London: Hutchinson & Co.).

37. This observation is made in L. Glage (1981), *Clementina Black: A Study in Social History and Literature* (Heidelberg: Winter), p. 42.

38. J. Goodman (1998), 'Social Investigation and Economic Empowerment: The Women's Industrial Council and the LCC Trade Schools for Girls, 1892–1914', *History of Education*, Vol. 27(3): 297–314 (307).

39. MacDonald, *Margaret Ethel MacDonald*, p. 137.

40. Ibid., p. 140.

41. Ibid., p. 131.

42. The tribute appeared in *Labour Leader*, 15 September 1911, and is quoted in M. A. Hamilton (1924), *Margaret Bondfield, 1873–1953* (London: L. Parsons), p. 78.

43. Hannam, 'MacDonald, Margaret Ethel Gladstone'.

9   Girls in Trades

1. K. Mansfield (1908), 'The Tiredness of Rosabel', in Richardson, *Women Who Did*, p. 337.

2. See Bondfield, *A Life's Work*.

3. P. Ackroyd (2000), *London: The Biography* (London: Chatto & Windus), p. 600.

4. T. Macnamara (1904), 'Physical Condition of Working-Class Children', *Nineteenth Century*, Vol. 56: 307–11 (308).

5. The novel was first published by Hodder & Stoughton in 1889 with the title *Captain Lobe: A Story of the Salvation Army*. Most slum novels were written by men and Harkness adopted the male pseudonym of 'John Law'.

6. M. Harkness (2003), *In Darkest London* (Cambridge: Black Apollo Press), p. 97.

7. 'What Becomes of Girls on Leaving School?' *WIN*, June 1900, p. 180.

8. 'A Report upon Technical Training for Women', *WIN*, No. 21, December 1902, p. 335 [the report was prepared by the Technical Training Committee of which Grace Oakeshott was Honorary Secretary and authorship is attributed to her on this basis].

9. Letter written by Grace Oakeshott, LSE, WIC/E/6.

10. 'A Report upon Technical Training for Women', p. 336.

11. G. Oakeshott (1904), 'The Need of Technical Education for Girls'. Paper delivered at a Conference of the National Union of Women Workers, 10 November, and published in *Conference Papers* (NUWW, 1904) which were consulted at the London School of Economics.

12. 'Technical Education for Women and Girls at Home and Abroad', WIC, 1904, p. 9 [the Education Committee of the Women's Industrial Council is named as corporate author of this report. Grace Oakeshott was Honorary Secretary of the relevant sub-committee and authorship is attributed to her on this basis].

13. 'Minutes of the Domestic Economy School Committee', the Educational Committee Minutes, London South Bank University (LSBU), LSBU/3/3/1.

14. LSBU, Borough Polytechnic Institute's Annual Reports. For 14th Annual Report (1904–5) see LSBU/6/8/13 and for 15th Annual Report (1905–6) see LSBU/6/8/14.

15. The syllabus for the Trade School for Girls at Borough Polytechnic Institute is attached to C. T. Millis (1909), 'Problems Connected with Trade Schools', an address given at Cardiff (30 March).

16. MacDonald, *Margaret Ethel MacDonald*, p. 141.

17. LSBU, Borough Polytechnic Institute's 16th Annual Report (1906–7), LSBU/6/8/17.

18. Board of Education (1924), 'Report of H.M. Inspectors on Junior Technical Education given in the London Trades Schools for Girls for the period ending 31st July 1924', TNA, ED 114/649.

19. The report on opportunities for girls in the photographic industry was drawn up from notes left by Grace and published in a compilation document, L.C.C. (1908) *Reports on Women's Trades* (London: LCC) after her presumed death.
20. See B. Bailey (1990), 'Technical Education and Secondary Schooling, 1905–1945', in P. Summerfield and E. Evans (eds), *Technical Education and the State since 1850: Historical and Contemporary Perspectives* (Manchester: Manchester University Press).
21. For an assessment of the role and impact of the Trade Schools for Girls, see J. Goodman (1998), 'Social Investigation and Economic Empowerment: The Women's Industrial Council and the LCC Trade Schools for Girls, 1892–1914', *History of Education*, Vol. 27(3): 297–314.
22. A. Bennett (1910), 'L.C.C. Technical Schools for Girls', *The Englishwoman*, Vol. 5(14): 219.
23. Tributes to Grace Oakeshott's work that appeared after her presumed death can be found in the Borough Polytechnic Institute's 16th Annual Report (1906–7), LSBU/6/8/17, and in L.C.C. (1908) *Reports on Women's Trades*, as well as in several WIC publications.
24. 'Further Report of the Education Committee', LCC Education Committee Minutes of Proceedings, March–April 1905. LMA, 22.06 LCC.
25. Grace's salary in 1905 is equivalent today to about £28,000 in terms of its purchasing power.
26. The Trade Schools were eventually swallowed up in further educational reforms.

10 Medical Men

1. *The Guyoscope*, 5 May 1897, p. 52.
2. Guy's Hospital Medical School student registers for 1894–1908, King's College London (KCL), G/FP9/2.
3. Unpublished family sailing logs.
4. Obituary, 'Henry James Cash', *Journal of Institution of Electrical Engineers*, July 1963, p. 317.
5. I am indebted to Bill Edwards, Curator at the Gordon Museum of Pathology, for his observations about both of the photographs in this chapter.
6. *Guy's Hospital Gazette*, 27 October 1900, pp. 501–2.
7. Unpublished family papers.
8. These terms are no longer in use at Guy's.
9. Building on work by Louis Pasteur on processes of fermentation, Joseph Lister became convinced in the early 1860s that invisible living particles were responsible for wound infections but interestingly, he himself never wore a mask, gown or gloves in the operating theatre.
10. Staff records for Guy's Medical School, LMA, H09/GY.
11. *The Guyoscope*, 5 May 1897, p. 70.
12. A. Conan Doyle [1894] (2007), 'His First Operation', in R. Darby (ed.), *Round the Red Lamp and Other Medical Writings* (Kansas City: Valancourt Books), pp. 12–13.
13. The bag probably contained ether.
14. Unpublished extract supplied by the Old Operating Theatre Museum, London.
15. Unpublished family papers.
16. Ibid.
17. Quoted in T. B. Layton (1956), *Sir William Arbuthnot Lane, Bt.: An Enquiry into the Mind and Influence of a Surgeon* (Edinburgh: E. & S. Livingstone), p. 105.

18. Ibid., p. 107.
19. G. B. Shaw [1906] (1946), *The Doctor's Dilemma* (London: Penguin Books), p. 18.
20. Shaw and fellow Fabians were critical of a tax that was to be imposed regardless of a person's ability to pay and wanted to see more far-reaching measures introduced to address poverty.
21. Unpublished family papers.
22. Layton, *Sir William Arbuthnot Lane*, p. 101.
23. Unpublished family papers.
24. Colonial Office records, TNA, CO429/29.
25. The reason the offer was withdrawn remains unclear. According to the family in New Zealand, when he applied for active service during the First World War, a heart murmur was detected. However, in the official records of his medical examination for military service, his heart was pronounced normal.

11   A Place to Begin Again

1. H. G. Wells [1909] (2005), *Ann Veronica* (London: Penguin Books), p. 256.
2. K. Summerscale (2012), *Mrs Robinson's Disgrace: The Private Diary of a Victorian Lady* (London: Bloomsbury). See Chapter 2 for a discussion of George Drysdale's case that draws on several primary sources.
3. The novel *Beautiful Lies* written by Clare Clarke (London: Harvill Secker, 2012) was inspired by this true story. Gabriela Cunninghame Graham's husband, Robert, was the first openly socialist MP and he held the seat for North-West Lanarkshire from 1886 to 1892.
4. John Wilson, 'The Voyage Out – Personal Accounts: 1900–1959', Te Ara, the Encyclopedia of New Zealand, available online at http://www.TeAra.govt.nz/en/community-contribution/4305/a-childs-voyage (accessed 5 March 2015).
5. Jock Phillips, 'History of Immigration', Te Ara, http://www.TeAra.govt.nz/en/history-of-immigration (accessed 20 January 2015).
6. A. Trollope (1873), *Australia and New Zealand*, Vol. III (Leipzig: Bernhard Tauchnitz), p. 216.
7. This community was never established but others were, and the country is still home to more 'intentional communities' per capita than any other in the world. Some members of the Clarion Fellowship were eventually involved in the formation of the New Zealand Socialist Party.
8. New Zealand became the first self-governing democracy to grant the right to vote to all adult women.
9. B. Webb (1898–1901), 'Diary of Our First Journey Round the World', Vols. 17–20, LSE, Passfield/1/ 2, p. 1760, entry for 3 August 1898.
10. Ibid., p. 1782, entry for 22 August 1898.
11. Ibid., p. 1773, entry for 3 August 1898.
12. Ibid., p. 1811, entry for 24 August 1898.
13. Mrs W. P. Reeves (1899), 'Women in New Zealand', *The Women's Industrial News* (March): 89–91.
14. Other British radicals who visited New Zealand around this time included Keir Hardie, Ben Tillett and Tom Mann.
15. R. Yska (2006), *Wellington: Biography of a City* (Auckland: Reed), and P. Lawlor (1959), *Old Wellington Days* (Wellington: Whitcombe & Tombs), p. 95.
16. B. Gustafson (1986), *From the Cradle to the Grave: A Biography of Michael Joseph Savage* (Auckland: Reed), p. 71.
17. 'Progress of the Fire', *The Evening Post (EP)*, 11 December 1907, p. 7.

18. 'Meeting of the Cabinet', *EP*, 11 December 1907, p. 8.
19. 'Parliament's Home Destroyed', *EP*, 11 December 1907, p. 7.
20. 'From the Tomb of a Great Man', *EP*, 11 December 1907, p. 7.
21. 'Ladies Column', *EP*, 14 December 1907, p. 19.

12 'Ignoble Motives'

1. 'Shooting Affair at Pakarae', *The Poverty Bay Herald (PBH)* 10 June 1908, p. 5.
2. The place where Walter was stranded is now a marine reserve. The modern spelling 'Tapuwae' means footprint in Maori and according to local tradition, the giant Rongokako left his footprint here, on the flat rocks, as he strode down the eastern seaboard of the North Island to escape an adversary.
3. See W. H. Oliver and J. M. Thomson (1971), *Challenge and Response: A Study of the Development of the East Coast Region* (Gisborne: The East Coast Development Research Association).
4. Monty Soutar, 'East Coast Places – Gisborne', Te Ara, http://www.TeAra.govt.nz/en/east-coast-places/page-6 (accessed 20 January 2015).
5. Oliver and Thomson, *Challenge and Response*, p. 152.
6. Ibid., p. 148.
7. Megan Cook, 'Women's Health – Pakeha Women's Health, 1840s to 1940s', Te Ara, http://www.TeAra.govt.nz/en/womens-health/page-3 (accessed 20 January 2015). Note that Maori were less fortunate. Colonisation brought them into contact with infectious diseases and, in addition, they suffered a loss of resources; wars also took their toll and by the 1890s Maori numbers had more than halved.
8. B. Webb (1898–1901), 'Diary of Our First Journey Round the World', Vols. 17–20, LSE, Passfield/1/2, p. 1811.
9. F. Porter and C. MacDonald (1996) (eds), *My Hand Will Write What My Heart Dictates* (Auckland: Auckland University Press/Bridget Williams Books), p. 455. Maternal mortality rates for Maori in 19th and early 20th centuries are unknown.
10. Ibid., p. 386.
11. Ibid., p. 347.
12. J. Mander (1920), *The Story of a New Zealand River* (London: John Lane), pp. 34–6.
13. Oliver and Thomson, *Challenge and Response*, p. 149.
14. Monty Soutar, 'East Coast Places'.
15. Personal communication from the historian, Elizabeth Cox.
16. This was the second Te Poho o Rawiri meeting house and it was built in about 1890. The first had been built close by on land that was taken for harbour works in 1886. The new site was permanently allocated to local Maori (for whom it had spiritual importance) but it, too, was taken for harbour works in 1929, an event that local people still feel wronged by. The third Poho o Rawiri, further away from the river but still on its eastern side, is now one of New Zealand's biggest carved meeting houses. I am indebted to Elizabeth Cox for these clarifications.
17. Porter and MacDonald, *My Hand Will Write*, p. 383.
18. Cook County was one of five counties in the East Coast region which were essentially political entities. The Cook County Women's Guild opened a day shelter for neglected children in 1908, built a maternity hospital in 1910, and erected a Children's Home in 1913.
19. L. Bryder (2003), *A Voice for Mothers: The Plunket Society and Infant Welfare 1907–2000* (Auckland: Auckland University Press), p. x [the figures did not include Maori whose infant mortality rate was not known].
20. Ibid., p. xi.

21. 'Science and the Child', *PBH*, 29 July 1912, p. 2.
22. 'For the Babies', *PBH*, 17 December 1913, p. 7.
23. Bryder, *A Voice for Mothers*, p. 44.
24. See E. Olssen (1981), 'Truby King and the Plunket Society. An Analysis of a Prescriptive Ideology', *The New Zealand Journal of History*, Vol. 15(1): 3–23.
25. Other factors such as a sharply declining birth rate, the extension of general education for girls, adequate food and lack of overcrowding also certainly contributed. The declining birth rate was linked in New Zealand (as in Britain) to concerns about the future of the race and the Empire. The Plunket Society was seen as 'patriotic' to the extent that it focused on the welfare of babies and children on whom the future depended. A minority in New Zealand at this time put forward eugenicist views, insisting that the weaker members of society should not be allowed to reproduce and that social policy should not pander to them, but more progressive voices insisted on the importance of the environment to child health and these views prevailed.
26. 'Children's Carnival', *PBH*, 30 August 1912, p. 5
27. 'Lamont Gurr Concert', *PBH*, 31 August 1912, p. 3.
28. Rosemary's sister, Annie Lee Rees, was a local teacher and she and Joan became close associates. Annie's and Rosemary's father, William Lee Rees, was a prominent lawyer and liberal politician (see Chapter 13, 'The Politics of Knitting').
29. K. Kelly (2004), '*Alan's Wife*: Mother Love and Theatrical Sociability in London of the 1890s', *Modernism/modernity*, Vol. 11(3): 539–60 (for a fuller account of the importance to socialists of the Marx/Aveling relationship, see Chapter 4, 'Another Word for Suicide').
30. See A. M. MacBriar (1966), *Fabian Socialism & English Politics 1884–1918* (Cambridge: Cambridge University Press), pp. 82–92.
31. Chapter 13, 'The Politics of Knitting', focuses on New Zealand's experience of the First World War and the couple's involvement in patriotic work.
32. Although nearly 60 per cent of the victims were under five years of age, significant numbers of adults were also affected.
33. E. Morris (n.d.), 'We're Still Here!', *Older Family Care New Zealand*, available online at http://www.catchword.co.nz/carers/fc_page1.html (accessed 20 January 2015).
34. New Zealand Defence Force personnel file for Walter Reeve, Archives New Zealand Te Rua Mahara o te Kāwanatanga (ANZ), R24268859.
35. 'Infantile Paralysis', *PBH*, 13 April 1916, p. 3. The English specialist Walter referred to was probably a Dr Robert Jones of Liverpool University, who in 1914 had described the importance of rest during acute and early convalescent stages of polio. Massage was to be avoided until all tenderness had receded. By 1916 electrotherapy was being widely used in America and Europe to treat polio victims but, along with massage, it would prove ineffectual in the long run. The breakthrough came with the development of effective vaccines in the 1950s.
36. 'Infantile Paralysis', *PBH*, 13 April 1916, p. 3.
37. 'Infantile Paralysis', *PBH*, 18 April 1916, p. 3.
38. Members of the hospital board were elected by local ratepayers.
39. 'Infantile Paralysis', *PBH*, 20 April 1916, p. 5.
40. 'Infantile Paralysis', *PBH*, 22 April 1916, p. 5.
41. 'Infantile Paralysis', *PBH*, 22 July 1916, p. 8.
42. A Royal Commission is a major formal inquiry set up by the state to investigate a matter of significant public interest. Within its terms of reference it has the power to summon witnesses who give evidence under oath. It can require the production of evidence and usually culminates in a report with policy recommendations.

43. 'Snubbed by the Board', *PBH*, 4 June 1917, p. 3.
44. 'Hospital Unrest', *PBH*, 26 July 1917, p. 6.
45. 'Hospital Commission', *PBH*, 18 January 1918, p. 4.
46. 'The Hospital Enquiry', *PBH*, 18 January 1918, p. 7.
47. D. Dow (1991), 'Springs of Charity? The Development of the New Zealand Hospital System, 1876–1910', in Linda Bryder (ed.), *A Healthy Country: Essays on the Social History of Medicine in New Zealand* (Wellington: Bridget Williams Books), pp. 44–64.
48. Ibid. Increasing numbers of private hospitals were established from the early 20th century. In 1908 there were 56 public hospitals in New Zealand and 191 private hospitals.
49. Mr Bishop's report appeared in full in the *Poverty Bay Herald* on 25 February 1918, 'Hospital Inquiry', p. 6ff.
50. I am indebted to the Gisborne historian, Marie Burgess, for this information.
51. 'Hospital Inquiry', *PBH*, 25 February 1918, p. 6.
52. 'Hospital Enquiry', *PBH*, 22 March 1918, p. 8.
53. 'Hospital Matters', *PBH*, 20 September 1918, p. 2.

13 The Politics of Knitting

1. Adapted from 'Untitled', *PBH*, 28 August 1914, p. 4. Joan would by now have read stories of German atrocities in Belgium. Modern scholars have found evidence of large-scale atrocities but the more sensational accounts have not been corroborated.
2. G. T. Chesney [1914] (2012), *The Battle of Dorking* (London: Forgotten Books), p. 52. The novel was first published in 1871 in a political journal and it soon triggered a spate of stories about hypothetical invasions. By the time war broke out in 1914 there were over four hundred such novels, and a worldwide readership.
3. T. Brooking (2004), *The History of New Zealand* (Connecticut and London: Greenwood Press), p. 98.
4. In areas of the country that had been aligned with the British Crown or colonial forces during the New Zealand Wars of the 1860s, young Maori were willing volunteers. There were far fewer from communities and tribes that had seen their land confiscated following those conflicts.
5. The title of the post was changed to 'Governor-General' in 1917.
6. The national parliament created four specifically Maori seats in 1867. M. King (2003), *The Penguin History of New Zealand* (Auckland: Penguin Books) notes that although this was a progressive move in some respects (New Zealand was the first neo-European country in the world to give the vote to its indigenous population, and Australia did not do so for another 95 years), if the seats had been allocated on the basis of population, as the non-Maori ones were, Maori would have had 14 or 15. It was not until near the end of the twentieth century that the number of Maori seats was at last increased and allocated on the basis of population.
7. 'Promenade Concert', *PBH*, 21 December 1914, p. 8.
8. 'Lady Liverpool's Scrapbooks', Vol. 4, September 1914–September 1915. National Library of New Zealand, Alexander Turnbull Library (ATL), P f920 LIV SCR 1912–1920.
9. Kerryn Pollock, 'Sewing, Knitting and Textile Crafts – Knitting, Spinning and Weaving', Te Ara, http://www.TeAra.govt.nz/en/sewing-knitting-and-textile-crafts/page-3 (accessed 20 January 2015).
10. 'Beautifying Work', *PBH*, 5 August 1914, p. 8.

11. B. Scates (2001), 'The Unknown Sock Knitter: Voluntary Work, Emotional Labour, Bereavement and the Great War', *Labour History*, No. 81, p. 39.
12. 'Lady Liverpool's Scrapbooks', p. 45.
13. The Australian and New Zealand Army Corps.
14. In the 1880s and 1890s, with Wi Pere, a local Maori leader, William Rees tried (ultimately unsuccessfully) to resolve some of the legal complexities of land ownership in the district, to ensure Maori owners received a fair price for their land and that sufficient was left for them to farm.
15. 'Cook County College', *PBH*, 18 December 1917, p. 7.
16. 'Untitled', *PBH*, 1 October 1915, p. 2.
17. 'Women's National Reserve', *Colonist*, 25 September 1915, p. 3.
18. 'National Reserve', *PBH*, 24 September 1915, p. 8.
19. J. Courage (1954), *The Young Have Secrets* (London: Jonathan Cape), p. 220.
20. The Defence Act introduced in 1909 required nearly all boys aged 12 to 14 to undertake 52 hours of physical training each year as Junior Cadets.
21. Courage, *The Young Have Secrets*, p. 214.
22. 'Camp Complaints', *PBH*, 9 March 1915, p. 7.
23. 'Equipment of the Soldiers', *PBH*, 13 March 1915, p. 4.
24. Ibid.
25. Scates, 'The Unknown Sock Knitter', p. 31.
26. Historians have come to see the Gallipoli campaign as a defining moment in the emergence of a distinct national identity for New Zealanders. Anzac Day (25 April) commemorates the landing at Gallipoli and the country's close ties with the Australians who fought alongside them.
27. Brooking, *The History of New Zealand*, p. 101.
28. M. McKinnon (1993) (ed.), *Bateman New Zealand Historical Atlas* (Auckland: Bateman, in association with New Zealand Department of Internal Affairs, Historical Branch), Plate 77.
29. 'Town Edition', *PBH*, 8 September 1916, p. 6.
30. 'Untitled', *PBH*, 10 August 1918, p. 2.
31. 'Untitled', *PBH*, 3 September 1918, p. 4.
32. 'Untitled', *PBH*, 17 September 1918, p. 2.
33. 'War Services', *PBH*, 7 October 1918, p. 5.
34. Mrs Pomare also received an OBE and Lady Liverpool was invested as a Dame Grand Cross of the Order of the British Empire (GBE). 'For War Services', *PBH*, 18 March 1918, p. 4.
35. 'Untitled', *PBH*, 19 October 1918, p. 2.
36. 'Gisborne Women's Patriotic Association', *PBH*, 4 August 1919, p. 5.
37. Figures here are taken from http://www.nzhistory.net.nz/war/first-world-war-overview/introduction (accessed 20 January 2015).
38. Brooking, *The History of New Zealand*, p. 103. Brooking notes the New Zealand casualty rate was second only to Australia's 65 per cent.
39. Ibid.
40. M. Nolan (2007), '"Keeping New Zealand's Home Fires Burning": Gender, Welfare and the First World War', in J. Crawford and I. McGibbon (eds), *New Zealand's Great War* (Auckland: Exisle Publishing), pp. 493–517 (515).
41. New Zealand Defence Force personnel file for Walter Reeve, ANZ, R24268859.
42. The most authoritative source on the New Zealand experience of the 1918 influenza pandemic is G. Rice (2005), *Black November: The 1918 Pandemic in New Zealand* (Christchurch: Canterbury University Press).

14   Landfall

1. T. Shoebridge (2011), *Featherston Military Training Camp and the First World War, 1915–27* (Wellington: Manatu Taonga/Ministry for Culture and Heritage), pp. 34–5.

2. Death rates were unevenly spread across the country and some other communities also escaped unscathed, particularly those that had received partial immunity from an earlier wave of the disease. Others were decimated.

3. The death rate amongst Maori was more than seven times that for the Pakeha population. In New Zealand the death rates for males were double those for females in the worst affected groups. Estimates of the number of people killed worldwide run from twenty-one to twenty-five million.

4. The letters from Walter and Dr Felkin are both contained in Walter's New Zealand Defence Force personnel file, ANZ, R24268859.

5. By the early 1900s, timber (which survives earthquakes far better than brick or stone) had become the dominant building material and corrugated iron was a popular choice for roofing. Detached houses set in their own 'section' were the ideal dwelling as far as most New Zealanders were concerned.

6. J. Siers (2007), *The Life and Times of James Walter Chapman-Taylor* (Napier: Millwood Heritage Productions), p. 195.

7. Robert Felkin had become acquainted with esoteric religious movements in Edinburgh in the 1880s. Following a spell in Africa as a medical missionary, he joined the Hermetic Order of the Golden Dawn, and then led an offshoot of that group called 'Stella Matutina'. This organisation had a particular interest in magic. In Havelock North, Felkin's purpose was to help the group 'Havelock Work' reach a higher spiritual plane.

8. Quoted in R. Ellwood (1993), *Islands of the Dawn: The Story of Alternative Spirituality in New Zealand* (Honolulu: University of Hawaii Press), p. 169.

9. Whare Ra was designed and built by the architect James Chapman-Taylor who was himself a member of the Hermetic Order of the Golden Dawn.

10. B. Anderson [1989] (1993), 'Poojah', in *I Think We Should Go Into The Jungle* (London: Secker & Warburg), pp. 37–8 [the writer explains that 'poojah' is a Sanskrit word for a ritual act of worship].

11. Ellwood, *Islands of the Dawn*.

12. Although its journal *The Forerunner* ceased publication in 1914, Havelock Work itself continued to exist through both world wars and survived until 1971, when land was purchased near Lake Taupo and the spiritual retreat 'Tauhara' was established. For an account of other spiritualist movements that found a home in Havelock North during the first decades of the twentieth century, including the Stella Matutina, Rudolf Steiner's Anthroposophical Society, and Robert Sutcliffe's School of Radiant Living, see M. Wright (1996), *Havelock North: The History of a Village* (Hastings: Hastings District Council), Chapter 11, 'Independent Thought'.

13. By this time primary schools in New Zealand were free, compulsory and secular. Private primary schools had to be registered and follow the same curriculum as state schools. Secondary education, which was expanding rapidly, did not become free for all pupils until after the election of the first Labour government in the 1930s, led by Michael Joseph Savage.

14. 'Overview', available online at http://www.nzhistory.net.nz/culture/the-1920s/overview (accessed 20 January 2015).

15. Ellwood, *Islands of the Dawn*, p. 4.

16. The Multiple Sclerosis Society, available online at http://www.nationalmssociety.org/ (accessed 20 January 2015).
17. *Collegian*, Wanganui Collegiate School magazine, December 1926, Issue 129, p. 8.
18. Unpublished family papers [Walter's letter was written in reply to a letter of condolence he had received from Malvina Lawson whose daughter, Margaret Noel Lawson, married his son, Tony, in 1934].
19. M. Fairburn (1989), *The Ideal Society and its Enemies: The Foundations of Modern New Zealand Society 1850–1900* (Auckland: Auckland University Press), p. 144.
20. *Hawke's Bay Tribune*, 12 December 1929, p. 4.

15   After Lives

1. Since 2009, in the UK, in the absence of a body, so long as there is some evidence to suggest the person has died, relatives can apply for a declaration of death without having to wait seven years.
2. Harold's thoughts on the ethics of marriage and the moral basis of socialism were set down in a series of unpublished family papers and have been paraphrased here.
3. G. Oxford (2006), 'The Prickly Willow Tree'. Unpublished family papers.
4. Ibid.
5. See D. Cohen (2013), *Family Secrets: Living with Shame from the Victorians to the Present Day* (London: Viking).
6. Both the Carr brothers became respected members of the scientific community. For 42 years Stanley was the General Secretary of the Chemical Society (which merged with several other organisations in 1980 to form the Royal Society of Chemistry), and Francis became a renowned chemist. Anna Carr was a talented musician. She was a prominent member of the Croydon Conservatoire of Music before emigrating to South Africa.
7. *Winterbourne Times 1907–2007: A Celebration of Winterbourne Schools – 100 Years with Memories Past and Present* (Croydon, London: Winterbourne Schools, 2007), p. 8.
8. London University first admitted women to its degrees in 1878. Cambridge University did not admit women to full university membership until 1947.
9. Kate Cash's registry file was consulted at the Records Office, University College London.
10. For details of Kate Cash's employment at the Froebel Educational Institute, see *FEI Committee Minute Books* for 1923 to 1929, UR, FA/FEI/2.2.
11. An obituary for Kate Cash was published in the *Herald for Farnham, Haslemere & Hindhead*, on 4 November 1949, p. 5. This mentions her interest in the suffrage movement but no detailed records of her political activities have been located.
12. 'Death of Dr Elizabeth Wilks', *Farnham Herald, Haslemere Herald, Alton Mail*, 23 November 1956, p. 8.
13. For more background about the settlement at Headley Down and Dr Wilks' involvement in it, see B. White (1999), 'The Influence of Dr Wilks on Headley', Headley Miscellany, Vol. 1 (December), pp. 5–8, available online at http://www.johnowen smith.co.uk/headley/vol1.htm#wilks (accessed 20 January 2015).
14. Records of Henry Cash's professional activities were consulted at the Electrical Contractors Association (ECA) and at the Institute of Engineering and Technology (IET), London.
15. Adapted from 'The Peace Dinner', *The Electrical Contractor*, November 1919, p. 291.

16. 'Henry James Cash', *Journal of the Institution of Electrical Engineers*, Vol. 9(103) (1963): 317.
17. Mason, *What Time is it now? My Life as a Teacher*, p. 24.
18. Dorothy had a difficult time giving birth to her first child, Harold Siegfried. She believed the boy's lifelong disabilities had been caused at his birth and this experience is what prompted her to train as a midwife.
19. Unpublished family correspondence.
20. Unpublished family correspondence.
21. 'Henry James Cash', *Journal of the Institution of Electrical Engineers*.
22. In the 1930s Renée was active in radical and socialist circles, keenly interested in art and literature and a member of the Left Book Club, a study and discussion group working for civil liberties and world peace. See R. Barrowman (1991), *A Popular Vision: The Arts and the Left in New Zealand 1930–1950* (Wellington: Victoria University Press). I am indebted to Rachel Barrowman for forwarding copies of her correspondence with Renée Stockwell (née Reeve).
23. Unpublished family correspondence.

Select Bibliography

I Unpublished Sources

Family Interviews and Papers

The families of Grace Cash, Harold Oakeshott and Walter Reeve have kindly shared their unpublished papers, photographs and memorabilia. Unless otherwise stated, all anecdotes and personal recollections are taken from recorded interviews and personal communications with family members between 2008 and 2014.

Archive Collections

Full archival references to material used in the chapters can be found in the Notes. Below is a key to abbreviations used.

Alexander Turnbull Library (ATL), National Library of New Zealand, Wellington.
Archives New Zealand Te Rua Mahara o te Kāwanatanga (ANZ), Wellington.
British and Foreign School Society Collection (BFSS), Stockwell College records, held at Brunel University, London.
City of Westminster Archive Centre (CWAC), London.
Croydon High School Archives (CHS), London.
Croydon Local Studies Library (CLSL), London.
Electrical Contractors Association (ECA), London.
Institute of Education Archives (IOE), University College London.
Institute of Engineering and Technology (IET), Stevenage.
King's College London (KCL), Archives and Special Collections.
London Metropolitan Archives (LMA).
London School of Economics (LSE), Archives and Special Collections.
London South Bank University (LSBU).
Minnesota Historical Society (MHS), St Paul, Minnesota.
Monkton Combe School Archives (MCS), Bath.
Royal Grammar School Newcastle (RGS).
The National Archive (TNA), Kew, London.
Thornton Heath Congregational Church Archives (CLSL).
University College London (UCL).
University of Birmingham (UBCRL), Cadbury Research Library.
University of Roehampton (UR), Archives and Special Collections.

II Published Sources

Key Newspapers and Periodicals

Englishwoman's Review
Newcastle Courant
Seed-time

Sunderland Daily Echo and Shipping Gazette
The Croydon Advertiser and Surrey County Reporter
The Croydon Times
The Evening Post (*EP*) (accessed online via the National Library of New Zealand)
The Poverty Bay Herald (*PBH*) (accessed online via the National Library of New Zealand)
The Times
The Women's Industrial News (*WIN*)

Websites

Where specific reference is made in the chapters to a website page, the full URL is given in the Notes. The most frequently consulted websites are listed below:

British History Online
   http://www.british-history.ac.uk
Encyclopaedia Britannica
   http://www.britannica.com
Measuring Worth
   http://www.measuringworth.com
National Library of New Zealand, Paperspast
   http://paperspast.natlib.govt.nz
New Zealand Electronic Text Collection
   http://www.nzetc.victoria.ac.nz
New Zealand Government, Department of Conservation
   http://www.doc.govt.nz
New Zealand History
   http://www.nzhistory.net.nz
Oxford Dictionary of National Biography
   http://www.oxforddnb.com
Te Ara, The Encyclopedia of New Zealand
   http://www.teara.govt.nz

Books and Academic Articles

Where specific reference is made to published texts in the chapters, full bibliographical details are given in the Notes. In addition, the following have provided valuable background material and are suggested as further reading.

Prologue

Koch, John T. (2006), *Celtic Culture: A Historical Encyclopaedia* (Santa Barbara: ABC-CLIO).

1   Radicals in Suburbia

Dyos, H. and M. Wolff (1973) (eds), *The Victorian City: Images and Realities* (Vol. 1) (London: Routledge & Kegan Paul).
Lewis, J. (1986) (ed.) *Labour and Love: Women's Experience of Home and Family, 1850–1940* (Oxford: Basil Blackwell).

McLeod, H. (1977), 'White Collar Values and the Role of Religion', in G. Crossick (ed.), *The Lower Middle Class in Britain 1870–1914* (London: Croom Helm), pp. 61–88.

Morris, J. N. (1992), *Religion and Urban Change: Croydon 1840–1914* (Woodbridge and Rochester, NY: The Boydell Press).

Morris, R. and R. Rodger (1993) (eds), *The Victorian City: A Reader in British Urban History 1820–1914* (London & New York: Longman).

Musgrove, F. (1959), 'Middle Class Education and Employment in the Nineteenth Century', *The Economic History Review*, Vol. 12(1): 99–111.

Thompson, F. (1988), *The Rise of Respectable Society: A Social History of Victorian Britain 1830–1900* (London: Fontana).

Watson, I. (1990), *Hackney and Stoke Newington Green Past* (London: Historical Publications).

2   Learning Curves

Allsobrook, D. (1986), *Schools for the Shires: The Reform of Middle-Class Education in Mid-Victorian England* (Manchester: Manchester University Press).

Brodie, J. and A. Laws (1924), *The Story of our School: The Royal Grammar School, Newcastle-upon-Tyne* (Newcastle upon Tyne: Northumberland Press).

Hyndman, M. (1978), *Schools and Schooling in England and Wales: A Documentary History* (London: Harper & Row).

Mains, B. and A. Tuck (1986), *Royal Grammar School Newcastle-upon-Tyne: A History of the School in its Community* (Stocksfield: Oriel Press).

Newton, S. (1974), *Health, Art and Reason: Dress Reformers of the 19th Century* (London: John Murray).

Purvis, J. (1991), *A History of Women's Education in England* (Milton Keynes: Open University Press).

Rapple, B. (1988), 'Matthew Arnold's Views on Modernity and a State System of Middle Class Education in England: Some Continental Influences', *The Journal of General Education*, Vol. 39(4): 206–21.

3   Finding Their Own Way

Dombkowski, K. (2003), 'Kindergarten Teacher Training in England and the United States, 1850–1918', *History of Education*, Vol. 32(1): 113–27.

Greenup, E. (1877), *Friendly Advice to Pupil Teachers* (London & Edinburgh: W & R Chambers).

Hanson, J. (1878), 'The Pupil Teacher System and the Instruction of Pupil Teachers', a paper read at a meeting of the Bradford and District Teachers' Association, 19 October (Bradford: Thomas Brear; London: Simpkin, Marshall & Co.).

Hunt, F. and C. Barker (1998), *Women at Cambridge: A Brief History* (Cambridge: Cambridge University Press).

Lupton, M. C. (1964), 'The Mosely Education Commission to the United States, 1903', *Journal of Vocational Education & Training*, Vol. 16(33): 36–49.

Perkin, J. (1993), *Victorian Women* (London: John Murray).

Read, J. (2003), 'Froebelian Women: Networking to Promote Professional Status and Educational Change in the Nineteenth Century', *History of Education*, Vol. 32(1): 17–33.

Weston, P. (2000), *Friedrich Froebel: His Life, Times and Significance* (Roehampton: University of Surrey).

4   'Another Word for Suicide'

Holmes, A. Sumner (1995), 'The Double Standard in the English Divorce Laws, 1857–1923', *Law & Social Inquiry*, Vol. 20(2): 601–20.

Hortsman, A. (1985), *Victorian Divorce* (London and Sydney: Croom Helm).

Jackson, J. (1969), *The Formation and Annulment of Marriages* (2nd edn.) (London: Butterworth).

Richardson, A. and C. Willis (2001) (eds), *The New Woman in Fiction and Fact: Fin-de-siècle Feminisms* (Basingstoke: Palgrave Macmillan).

Rowbotham, S. (2010), *Dreamers of a New Day: Women who Invented the Twentieth Century* (London and New York: Verso).

Savage, G. (1983), 'The Operation of the 1857 Divorce Act, 1860–1910: A Research Note', *Journal of Social History*, Vol. 16(4): 103–10.

5   'Fellowship is Heaven'

Armytage, W. (1961), *Heavens Below: Utopian Experiments in England, 1560–1960* (London: Routledge & Kegan Paul).

Britain, I. (1982), *Fabianism and Culture: A Study in British Socialism and the Arts c.1884–1918* (Cambridge and New York: Cambridge University Press).

Darley, G. (2007), *Villages of Vision: A Study of Strange Utopias* (Nottingham: Five Leaves).

Greenslade, W. (2007), 'Socialism and Radicalism', in G. Marshall (ed.), *The Cambridge Companion to the Fin de Siècle* (Cambridge: Cambridge University Press), pp. 73–90.

Hardy, D. (1979), *Alternative Communities in Nineteenth Century England* (London: Longman).

MacKenzie, N. (1979), 'Percival Chubb and the Founding of the Fabian Society', *Victorian Studies*, Vol. 23(1): 29–55.

Manton, K. (2003), 'The Fellowship of the New Life: English Ethical Socialism Reconsidered', *History of Political Thought*, Vol. 24(2): 282–304.

Morris, J. N. (1992), *Religion and Urban Change: Croydon 1840–1914* (Woodbridge and Rochester, NY: The Boydell Press).

Pease, E. (1916), *History of the Fabian Society: The Origins of English Socialism* (St Petersburg: Red and Black Publishers).

Yeo, S. (1977), 'A New Life: The Religion of Socialism in Britain, 1883–1896', *History Workshop*, No. 4: 5–56.

6   Answering the Call

Kelcey, B. (2001), *Alone in Silence: European Women in the Canadian North before 1940* (Montreal and Kingston, London, Ithaca: McGill-Queen's University Press).

McCormack, P. (2010), *Fort Chipewyan and the Shaping of Canadian History 1788–1920s* (Vancouver: University of British Columbia Press).

Mockridge, C. (1896), *The Bishops of the Church of England in Canada and Newfoundland* (Toronto: F. N. W. Brown).

Russell F. (1898), *Explorations in the Far North: Being the Report of an Expedition under the Auspices of the University of Iowa during the years 1892, '93 and '94* (Iowa City: University of Iowa).

7 Not Much 'Home' About It

Hughes, M. V. (1934), *A London Child of the 1870s* (Oxford: Oxford University Press).
Peters, L. (2000), *Orphan Texts: Victorian Orphans, Culture and Empire* (Manchester and New York: Manchester University Press).
Stock, E. (1899), *History of the Church Missionary Society: Its Environment, Its Men, and Its Work*, Vols. 1–4 (London: Church Missionary Society).
Wilson, A. H. (2007), *The Victorians* (London: Hutchinson).

8 Behind Closed Doors

Crawford, E. (1999), *The Women's Suffrage Movement: A Reference Guide, 1866–1928* (London: UCL Press).
Feldman, D. and G. Stedman Jones (1989) (eds), *Metropolis London: Histories and Representations since 1800* (London and New York: Routledge).
Hilton, M. and P. Hirsch (2000), *Practical Visionaries: Women, Education and Social Progress 1790–1930* (Harlow: Pearson Education).
Ledger, S. and R. Luckhurst (2000), *The Fin de Siècle: A Reader in Cultural History, c.1880–1900* (Oxford: Oxford University Press).
Reeves, M. P. (1979), *Round About a Pound a Week* (London: Virago).
Rowbotham, S. (2010), *Dreamers of a New Day: Women who Invented the Twentieth Century* (London and New York: Verso).

9 Girls in Trades

Bayley, E. (1910), *The Borough Polytechnic Institute: Its Origin and Development* (London: Elliot Stock).
Gibbon, G. and R. Bell (1939), *History of the London County Council, 1889–1939* (London: Macmillan).
Griggs, C. (1983), *The Trades Union Congress & the Struggle for Education, 1868–1925* (Lewes: Falmer).
Pennybacker, S. (1995), *A Vision for London 1889–1914: Labour, Everyday Life and the LCC Experiment* (London and New York: Routledge).
Purvis, J. (1989), *Hard Lessons: The Lives and Education of Working-Class Women in Nineteenth-Century England* (Cambridge: Polity Press).
Schneer, J. (1999), *London 1900: The Imperial Metropolis* (New Haven, CT and London: Yale University Press).
Simon, B. (1965), *Studies in the History of Education: Education and the Labour Movement 1870–1920* (London: Lawrence & Wishart).
Thompson, P. (1967), *Socialists, Liberals and Labour: The Struggle for London 1885–1914* (London and Toronto: Routledge).
Tomalin, C. (1987), *Katherine Mansfield: A Secret Life* (London: Viking).

10 Medical Men

Bonner, T. N. (1995), *Becoming A Physician: Medical Education in Britain, France, Germany and the United States, 1750–1945* (New York and Oxford: Oxford University Press).
Cameron, H. C. (1954), *Mr Guy's Hospital, 1726–1948* (London: Longmans, Green & Co.).

Crozier, A. (2007), *Practising Colonial Medicine: The Colonial Service in East Africa* (London: I. B. Tauris).

Dally, A. (1996), *Fantasy Surgery, 1880–1930: With Special Reference to Sir William Arbuthnot Lane* (Amsterdam and Atlanta, GA: Editions Rodopi B.V.).

Darby, R. (2007), 'Introduction', in A. Conan Doyle [1894], *Round the Red Lamp and Other Medical Writings* (Kansas City: Valancourt Books).

McBriar, A. M. (1966), *Fabian Socialism and English Politics, 1884–1918* (Cambridge: Cambridge University Press).

Peterson, M. J. (1978), *The Medical Profession in Mid-Victorian London* (Berkeley, Los Angeles and London: University of California Press).

Waddington, K. (2003), *Medical Education at St Bartholomew's Hospital, 1123–1995* (Woodbridge: The Boydell Press).

Witte, W. and C. Stein (2010), 'Chapter 1: History, Definitions and Contemporary Viewpoints', in A. Kopf and N. B. Patel (eds), *Guide to Pain Management in Low-Resource Settings* (Seattle: IASP).

11 A Place to Begin Again

Adams, J. (2007), 'Gabriela Cunninghame Graham: Deception and Achievement in the 1890s', *English Literature in Transition 1880–1920*, Vol. 50(3): 251–68.

Alessio, D. (2008), 'Promoting Paradise: Utopianism and National Identity in New Zealand, 1870–1930', *New Zealand Journal of History*, Vol. 42(1): 22–41.

Belich, J. (2001), *Paradise Reforged: A History of the New Zealanders from the 1880s to the Year 2000* (London: Allen Lane).

Fairburn, M. (1989), *The Ideal Society and its Enemies: The Foundations of Modern New Zealand Society 1850–1900* (Auckland: Auckland University Press).

MacKinnon, M. (ed.) with B. Bradley and R. Kirkpatrick (1997), *New Zealand Historical Atlas* (Auckland: Bateman/Department of Internal Affairs).

Sargisson, L. and L. T. Sargent (2004), *Living in Utopia: New Zealand's Intentional Communities* (Aldershot: Ashgate).

Simpson, T. (1997), *The Immigrants: The Great Migration from Britain to New Zealand, 1830–1890* (Auckland: Godwit).

12 'Ignoble Motives'

Mackay, J. A. (1949), *Historic Poverty Bay and the East Coast, N.I., N.Z.* (Gisborne, NZ: J. A. Mackay).

Ross, J. (1993), *A History of Poliomyelitis in New Zealand* (MA thesis, University of Canterbury, New Zealand).

Tennant, M. (1989), *Paupers and Providers: Charitable Aid in New Zealand* (Wellington: Allen & Unwin).

Williams, G. (2013), *Paralysed with Fear: The Story of Polio* (Basingstoke: Palgrave Macmillan).

13 The Politics of Knitting

Coney, S. (1993), *Standing in the Sunshine: A History of New Zealand Women since they Won the Vote* (Auckland: Viking).

Else, A. (1993), *Women Together: A History of Women's Organisations in New Zealand* (Wellington: Department of Internal Affairs).

Graham, J. (2008), 'Young New Zealanders and the Great War: Exploring the Impact Legacy of the First World War, 1914–2014', *Pedagogica Historica*, Vol. 44(4): 429–44.

Hobsbawm, E. (1987), *The Age of Empire, 1875–1914* (London: Weidenfeld & Nicolson).

Hucker, G. (2009), '"The Great Wave of Enthusiasm": New Zealand Reactions to the First World War in August 1914 – A Reassessment', *New Zealand Journal of History*, Vol. 43(1): 59–75.

Hutching, M. (2007), 'The Moloch of War: New Zealand Women Who Opposed the War', in J. Crawford and I. McGibbon (eds), *New Zealand's Great War* (Auckland: Exisle Publishing).

Oliver, W. H. and J. M. Thomson (1971), *Challenge and Response: A Study of the Development of the East Coast Region* (Gisborne, NZ: The East Coast Development Research Association).

Olssen, E. (1981), 'Towards a New Society', in W. Oliver (ed.) with B. Williams, *The Oxford History of New Zealand* (Oxford and Wellington: The Clarendon Press and Oxford University Press).

Walker, R. (1990), *Ka Whawhai Tonu Matou: Struggle Without End* (Auckland: Penguin Books).

14   Landfall

Brooking, T. (2004), *The History of New Zealand* (Connecticut and London: Greenwood Press).

Grant, S. (1978), *Havelock North: From Village to Borough, 1860–1952* (Hastings: Hawkes Bay Newspapers).

Sargisson, L. and L. T. Sargent (2004), *Living in Utopia: New Zealand's Intentional Communities* (Aldershot: Ashgate).

White, G. (2014), *Dr Robert Felkin: Magician on the Borderland* (Napier: MTG Hawkes Bay).

15   After Lives

Frost, G. (2008), *Living in Sin: Cohabiting as Man and Wife in Nineteenth Century England* (Manchester and New York: Manchester University Press).

Perkin, J. (1993), *Victorian Women* (London: John Murray).

Index

activism   16–17, 41, 54, 113, 152, 120
  see also Fabian Society;
  Fellowship of the New Life *and*
  Jupp, Rev. William
Adams, Maurice   18, 54–5, 57, 64, 204
  n7, n9 *see also* activism; Jupp,
  Rev. William *and* Fellowship of
  the New Life
adultery   50, 51, 203 n23 *see also*
  cohabitation *and* marriage
Anglican missions   69, 70, 76, 78, 80, 81,
  82, 83–4, 88–9, 206 n9
  see also Bompas, William; Church
  Missionary Society *and* Reeve,
  William Day
Arnold, Matthew   19–20
Arzon   4–5, 137, 197
asylums   11, 45
Aveling, Edward   44–5, 154
  see also socialism

Babbacombe Bay   2, 60–1
birth control   49–50, 214 n25
Black, Clementina   46, 97–8, 100, 103,
  208 n2 *see also* suffrage movement
  and Women's Industrial Council
boarding school   25–6, 94, 180, 183
  see also Monkton Combe School
  and Townley House Ladies School
Bompas, Selina   74, 76–9, 81
Bompas, William   73–6, 83–4, 206 n19
  see also evangelism
  Bishop of Athabasca   77, 81, 206 n18
Bondfield, Margaret   98, 107, 109,
  110–11 *see also* Women's
  Industrial Council
Borough Polytechnic Institute   107,
  115, **116**, 117, **118** *see also* Trade
  Schools
Brittany   2, 4, 133, 137, 150, 191
Burnham-on-Crouch   2, 60, 63

Caird, Mona   46 *see also* suffrage
  movement

Cambridge University   23, 202 n22, 218
  n8 *see also* Newnham College
Canada   73–4, 78, 80, 85, 124, 132, 153
  see also colonial life; Fort Rae;
  Fort Simpson *and* Hudson's Bay
  Company
  activities of missionaries *see* Anglican
    missions; Bompas, William;
    Reeve, William Day *and* Roman
    Catholic missions
  indigenous peoples   69, 74, 77–8,
    79–80, 81–2, 83–4, 85 *see also*
    Dene peoples
  Northwest Territories as part of   69, 72,
    73, 132, 206 n10
Canon Bob *see* Hall, Rev. Alfred
Cash, Elizabeth   6, 9, **10**, 11, 14, 15, 17,
  19, 29–30, 31, 151, 184, 189
Cash, Henry James   9, **10**, 123–4, 188–9,
  193, **194**, 197
  education   20–1
  yachtsman   53, 60–4, 205 n34, n35,
    n36 *see also* Oakeshott, Grace
    Marion, meeting Walter *and*
    staged drowning
Cash, James   6, 9, 11, 13–15, 18, 19, 24–5,
  29, 30, 31, 34, 55, 60, 184, 189
Cash, Jessie Elizabeth   9, **10**, 17, 18, 43,
  188–9, 190, 191
  education   17, 24–6, 30 31–4
  headmistress
    at Sydenham Road Girls'
      School   34–5, 42, 202 n12
    at Winterbourne Junior Girls'
      School   190–1
Cash, Kate Gertrude   **10**, 188–9, 191–3, 218
  n11 *see also* Wilks, Dr Elizabeth
  education   24–6, 30 *see also* Townley
    House Ladies School
  teaching career   34, 35–6, 42–3, 202
    n15 *see also* kindergarten
childbirth   75, 80, 148–9, 151, 190, 219
  n18 *see also* midwifery
  infanticide practices   76, 80

CHSG *see* Croydon High School for Girls
Church Missionary Society   63, 69–72, 77,
      81, 82–5, 87–91, 93, 94, 124, 175
Clough, Anne Jemima   37–9
      see also Newnham College
CMS *see* Church Missionary Society
cohabitation   44–5, 51–2
colonial life   139, 143, 146–7, 150–1,
      177, 196
   in New Zealand   139–40, 141, 142–3,
      147–8, 149
colonial medical service   122, 132–3
      see also Reeve, Walter
colonialism   132, 139, 215 n4
Cook County College for Girls   167–8, 170
Corke, Helen   13, 33–5, 59
Coulsdon   2, **8**, 16, 34, 47, **48**, 59, 64,
      111, 124, 186, 189
Courtauld, Renée   40–2
Croydon   5, **8**, 9, 11, 16–17, 55, 58–9, 65,
      124, 133, 147, 189–91, 195, 200
      n23, 204 n7, n15
Croydon High School for Girls   26–30,
      32, 36–7, 89, 117, 201 n26
      see also Neligan, Dorinda
Croydon Kindergarten Preparatory School
      Co. Ltd   35, 36
Croydon Times, The   186–7

Dene peoples   74–6, 78–80, 82–3, 206
      n10 *see also* Anglican missions
      and Roman Catholic missions
Dingemans, Sophie   xiii, 1
disability, stigma of   189, 193–4
disappearances   138–9
divorce   44, 50, 51–2, 139, 187, 203 n23
      see also marriage
Dorcas Society   17, 98
Downside Cottage   47, 48, 59–60, 64,
      100, 111, 120, 138, 203 n15

education   11, 18, 24–7, 29–32, 34–6, 38,
      41, 43–4, 88, 89, 92, 111–12, 115,
      117, 180, 190–1, 200 n1,
      214 n25, 217 n13 *see also*
      evangelical education; fee-paying
      education; secondary education;
      technical education *and* Women's
      Industrial Council
education of girls   11, 23, 24–9, 30, 88–9,
      92, 100–1, 113–19 131, 152, 167,
168, 170 *see also* Girls' Public Day
      School Company; kindergarten;
      Neligan, Dorinda; Newnham
      College; Stockwell Training
      College; technical education;
      Townley House; Trade Schools
      and Women's Industrial Council
   factors affecting   25–6, 106, 112–14, 118
      see also London County Council
   pupil teachers   31–3, 34
   teaching careers   35–6, 37, 38, 41–2,
      43, 190–1
educational reform   19–21, 24–6, 35, 42
      see also Arnold, Matthew *and*
      kindergarten
   Taunton Report   20–1, 23, 25
elementary school   31–4, 36, 112, 114,
      115, 117, 120, 168, 200 n1
endowed schools   20–1, 25, 42
equality   18, 51, 120, 152, 191, 192
   religious   16
   social   43
evangelical education   70–1, 72, 94–6, 123
      see also Church Missionary Society
      and Monkton Combe School
evangelism   63, 76, 77, 82, 83–4, 190,
      208 n18
evening courses   23–4, 33, 56–7,
      113–14, 122

Fabian Society   24, 44, 49, 51, 54, 131,
      141, 154, 162–3, 187, 200 n11,
      204 n6, 205 n27, 208 n2, 212 n20
Fawcett, Millicent   38, 120
      see also suffrage movement
fee-paying education   20–1, 25–6, 31,
      33–4, 37, 115–16, 125, 200 n2
Felkin, Dr Robert   177–80, 217 n7
      see also Havelock Work *and*
      influenza epidemic
Fellowship of the New Life   17–18, 24,
      44, 46, 48, 52, 53–8, 61, 64, 65,
      98, 179, 204 n9 *see also* Jupp,
      Rev. William *and Seed-time*
finishing school   26
First World War   161, 172–3, 174, 191,
      193, 212 n25 *see also* influenza
      epidemic
   expeditionary force   164, 175
   fundraising for   154, 162, 164–5, 168,
      170–2

hospital ships 143, 165
knitting for 165–6, 168, **169**, 170
medical work 129, 164
pacifism 164
patriotism 162–4, 165, 166, 167–8
soldiers' club 165, 168, 173
Women's Patriotic Committee 162,
165, 171–2, 173–4
Fort Rae 79–82, 85 *see also* Anglican
church *and* Hudson's Bay
Company
Fort Simpson **68**, 73–7, 79, 81–3, 85, 206
n9 *see also* Anglican church *and*
Hudson's Bay Company
freedom 18, 45, 49, 51, 54, 55, 151, 187
Froebel, Friedrich 35–6, 192
see also kindergarten

Girls' Public Day School Company 11,
25–6, 30, 31, 35, 36, 89, 199 n4 *see
also* Croydon High School for Girls
Gisborne 41, 145–8, 149, 150, 152–4,
155–61, 162, 164–6, 167–8,
170–2, 173–5, 177, 179, 180,
182, 196
GPDSC *see* Girls' Public Day School
Company
Guy's Medical School 122–32, 157 *see
also* Lane, William Arbuthnot *and*
Reeve, Walter

Hall, Kitty 196–7
Hall, Rev. Alfred 196–7
Havelock North 93, 176–8, 180–1, 183,
184, **185**, 196–7, 217 n7, n12
Havelock Work 178–80, 217 n12
Hawkes Bay 174, 176–7, 180, 182
HBC *see* Hudson's Bay Company
Hermetic Order of the Golden
Dawn 178, 217 n7, n9
Hicks, Amie 46, 98, 100, 103, 109 *see
also* Women's Industrial Council
high school 29, 32, 35–6 *see also*
Croydon High School for Girls
hospital, Gisborne 155–8
Hudson's Bay Company 74, 78–80

illnesses *see also* Church Missionary
Society *and* influenza epidemic
diphtheria 93–4, 123–4, 175
dysentery 153

measles 91
mental illness 34–5, 47
multiple sclerosis 183
poliomyelitis 155–7, 175, 214 n35
rheumatoid arthritis 82, 124, 132
scarlet fever 40, 76, 91, 123
typhoid 91, 123, 175
ILP *see* Independent Labour Party
imperialism 72, 111, 123, 163–4
Independent Labour Party 5, 51–2,
162, 190
infant welfare 111–12
influenza epidemic 155, 174–5, 176–8

Jupp, Reverend William 17, 34, 54, 55,
200 n23, 204 n7 *see also* activism
and Fellowship of the New Life

Kaiti 149, 150, 151, 168–70
kindergarten 11, 34, 35–6, 42, 55–6,
152, 167, 170, 191–3, 202 n20
King, Frederic Truby 151–4 *see also*
Plunket Society
knitting *see* First World War
Knox, Rev. F. V. 92–3

Lane, William Arbuthnot 128–32
Lawrence D. H. 13, 33, 59, 199 n10
LCC *see* London County Council
learning the truth 126, 197–8
Lees, Edith 58–9, 205 n27 *see also* Fabian
Society *and* New Woman
Liverpool, Lady 164–6, 171, 216 n34
London County Council 42, 99, 107,
119–21, 187, 202 n12 *see also*
Borough Polytechnic Institute;
Trade Schools *and* waistcoat-making
Grace's investigations for 113–16,
211 n19

MacDonald, J. Ramsay 100–1, 108–9,
121, 141, 162, 204 n5 *see also*
Fellowship of the New Life
MacDonald, Margaret 100–1, 103, 104,
109, 114, 117–18, 141, 187, 208
n3 *see also* Women's Industrial
Council
Maori 140, 149–50, 182, 213 n2, n9, 215
n6, 217 n3
effects of colonization on 213 n7,
n16, 216 n14

Maori – *continued*
 Pomare, Miria 165, 216 n34
 schooling 168
 as soldiers 164, 165, 215 n4
 voting rights 174, 215 n6
marriage 17, 43–52, 105, 113, 133, 138,
  151, 186–7, 193, 196, 203 n1,
  218 n2 *see also* social convention
  and respectability
 bigamous 187
 companionate 17, 50
Marx, Eleanor 44–5, 154, 205 n27
 see also socialism
medical advances 124, 126–7, 128–9, 130
medical controversies 130–2, 155–61
medical facilities, lack of 80, 149
Michaelis, Emilie 35–6, 38, 191
 see also kindergarten
midwifery 80, 149, 195, 219 n18
migration 98, 111, 132–3 *see also*
  colonial life *and* missionary life
 favourable views on 133, 141,
  148, 149
 Great Migration 139
 realities of 142, 143, 147–8, 151,
  153–4, 181–2 *see also* hospital,
  Gisborne *and* Plunket Society
 reasons for 139–40, 151
missionary life 72–4, 76, 84, 87, 89–90
 see also Dene peoples
 hardships of 75, 77–9, 80
Monkton Combe School 94, **95**, 122–3
Morris, William 44, 50–1, 53

Neligan, Dorinda 26–7, **28**, 36, 38, 41,
  46, 201 n17, n24, 202 n20 *see
  also* Croydon High School for
  Girls *and* kindergarten
New Woman 9, 47, 58, 106
New Zealand *see also* colonial life; First
  World War; Gisborne; Havelock
  North; Maori; Pakeha; Poverty
  Bay *and* Wellington
 as loyal dominion 142, 149, 163–4, 174
 migration to 132–3, 139–40, 149
 national mood change 181–2
 role in First World War 163–6, 167,
  168, 170–2, 174, 216 n26, n38
 student recruitment from 123, 140
 as utopia 137, 140, 182

Newnham College 30, 37–41
Nonconformism 16–18, 24, 29, 43, 55, 61,
  63, 88, 122, 200 n18, n19 *see also*
  activism; Fellowship of the New
  Life; London Missionary Society
  and Stockwell Training College

Oakeshott, Dorothy Frances (née
  Silverlock) 5, 186, 187, **188**,
  189, 193, 194–5, 219 n18
 children of 193–5, 219 n18
Oakeshott Eliza Maria (née Dodd) 17,
  21, 23, 64, 203 n15
Oakeshott, Grace Marion (née Cash)
  see also Reeve, Joan Leslie *and*
  staged drowning
 childhood **10**, 11–14
 education 11, 24–6, 27, **28**, **29**, 30,
  37, 38–41 *see also* Croydon High
  School for Girls; Girls' Public Day
  School Company *and* Newnham
  College
 erasure from records 189
 marriage to Harold 43–6, **48**, 50
  see also Downside Cottage *and*
  Ranmore Common
 meeting Walter 61–5, 126
  see also Ranmore Common
 obituary 5, 186
 as social activist/reformer 46, 47–8,
  57, 97–102, 104–9, 111–12
  see also Fellowship of the New
  Life; London County Council *and*
  Women's Industrial Council
 working life
  as inspector for women's technical
  classes 119–21
  for the LCC 42, 113–19
  in schools 41–2, 46
Oakeshott, Harold Augustus 17, **22**, **65**,
  188, 189, 193–6, 203 n15 *see also*
  Fabian Society; Fellowship of the
  New Life; Independent Labour
  Party *and* staged drowning
 activism 24, 46, 47, 48–50, 51–2, 55–6,
  58, 59, 162–3, 179, 200 n11
 Fellowship Gild, work for 56–7
 and alcohol 63, 64, 65, 194
 education 21–4, *see also* Royal
  Grammar School

marriage to Grace 41, 43, 44, 45, 46, 50
sailing 61, 63, 64
second marriage of *see* Oakeshott,
   Dorothy Frances
siblings of 17, 23, 24, 35, 55, 64, 203
   n15 *see also* Oakeshott, Joseph
   Francis
Oakeshott, Joseph 17, 21, 23–4, 203 n15
Oakeshott, Joseph Francis 17–18, 24, 51,
   53–4, 55, 59, 131, 141, 200 n11,
   204 n6 *see also* Fabian Society *and*
   Fellowship of the New Life

Pakeha 148–51, 165, 168, 174, 182,
   217 n3
Parliament Buildings fire 143–4
Plunket Society 146, 152–4, 164, 172,
   182, 214 n25
political communities 17, 55–6, 59 *see*
   also socialism *and* Fabian Society
political ideology 16, 18, 40, 46, 48, 141
   see also Fellowship of the New Life
political reform 47, 99, 107–8, 121, 140
Poverty Bay 145, 165, 168
Poverty Bay Herald 156, 157, 158
progressive outlook 22, 50, 55, 113, 120,
   121, 187, 214 n25, 215 n6

radicalism 16, 43, 47, 55, 120–1, 122,
   139–40, 141, 143, 187, 212 n14,
   219 n22 *see also* Fellowship of the
   New Life *and* Nonconformism
Ranmore Common 59, 64, 147
Ranmore (house) 178, 179, 181, 182
real drownings 5
Rees, Annie Lee 165, 167–8, 170, 214
   n28 *see also* Cook County College
   for Girls
Reeve, Antony Walter 148, 151, **152**,
   161, 168–70, **173**, 179, 180,
   182, 183, 196, 218 n18 *see also*
   learning the truth
Reeve, Colin David 148, **152**, 168–70,
   173, 179, 180, 183, 196 *see also*
   learning the truth
Reeve, Emily (née Parker) 69, 72, 74,
   75, 77–80, 82–3, 84, 85, 89, 124,
   132, 190
Reeve, Herbert Charles 77, 82, 87, 88,
   95–6, 125, 133, 190

Reeve, Joan Leslie ('née Knight')
   arrival in Wellington 141–5
   death of 183–5
   family life in Havelock North 176–8,
      180–1, 182
   fear of discovery 182–3
   Girl Guide Company 181
   journeying abroad 137–41
   life in Gisborne
      social involvement, Gisborne 151–5,
         164–7, 170–2 *see also* Plunkett
         Society
      Cook County College for
        Girls **167**, 168, 170
      Cook County Women's Guild 151,
        213 n18
      MBE award 173
      Poverty Bay Women's National
        Reserve 168
      soldiers' club 165, 168, 173
      Women's Patriotic Committee 162,
        165, 171–2, 173–4
   motherhood 148–9, 151
   multiple sclerosis 183
   new life 140, 141, 142, 143–4, 147–50
   Women's Institute 181
Reeve, Renée Mary Elizabeth 41, 151,
   154–5, 155, 167, **173**, 179, 180,
   183, 196 *see also* learning
   the truth
Reeve, Walter 69, **70**, 79, 81, 82, **95**,
   132, 137, 139–40, 142, 148–50,
   173, 182, 213 n2 *see also* Church
   Missionary Society; Guy's
   Medical School; Hall, Kitty *and*
   Monkton Combe School
   as actor 154, 167
   childhood 78–9
   death of Joan 183–5, 189–90
   education 72, 85, 88, 90, 91, 92, 94
   and Henry Cash 60, 205 n35
   Islington Home (CMS) 85–7, 88, 90, 91
   Limpsfield Home (CMS) 91–3
   locum in Wellington 143
   medical practice in Gisborne 146,
      151, 153, 155–61 *see also* medical
      controversies; Scott, Dr C. F. *and*
      Tait, Matron
   medical practice in Havelock
      North 178–80, 214 n35

Reeve, Walter – *continued*
  meeting Grace   61, 63, 64, **65**, 163, 186
  National Reserve   164, 168, 173
  siblings of   60, 77, 82, 87, 89, 190
    see also Reeve, Herbert Charles
  war call-up   172, 174–5, 176–7
    see also influenza epidemic
Reeve, William Day   69–70, 72–6, 77–82,
    85, 89 *see also* Bompas, William;
    Church Missionary Society;
    evangelism *and* missionary life
  Bishop of Mackenzie River   83, **84**,
    123, 206 n18
reform   158–61, 179, 182
  economic   54 *see also* Fabian Society
  educational   19–20, 24–25, 35–6, 211
    n26 *see also* Arnold, Matthew *and*
    Taunton Report
  employment   98–100 *see also* Women's
    Industrial Council
  social   44, 47, 55, 59, 100, 141
religion   16, 24, 29, 55, 63, 92 *see also*
    Anglican missions; Church
    Missionary Society; missionary
    life; Nonconformism *and* Roman
    Catholic missions
respectability   15–16, 18, 31, 36, 40,
    43, 46, 50–2, 64, 132, 133, 144,
    187–90
RGS   *see* Royal Grammar School
role of women   29, 39, 43, 47–9, 72,
    77–9, 108, 120, 121, 141, 152–3,
    164, 172–4, 181
Roman Catholic missions   76, 78, 81–3
Royal Grammar School   21–3

St David's Church   74, **75**
scholarships   33, 34, 94, 112, 114, 120–1,
    191, 200 n1
school inspectors   19, 42, 110, 119–20
Scott, Dr C. F.   155–7, 160
secondary education   19–20, 21, 24–6,
    30, 31, 36, 112, 117, 119–20, 200
    n1, 217 n13
Seed-time   53–4, 204 n9
sex   44, 51–2, 138–9
  abstinence from   49–50
    see also birth control
Shaw, George Bernard   49–50, 131, 154,
    212, n20 *see also* Fabian Society

social convention   15, 48, 57–8, 137
  see also respectability
  rejection of   44–5, 52, 63, 111, 138,
    154, 196
socialism   16, 17, 24, 44–6, 49–52, 53,
    55, 59, 139–40, 154, 162, 182,
    187, 200 n11, 204 n5, n6, 212
    n3, 218 n2, 219 n22 *see also*
    Fabian Society *and* Fellowship of
    the New Life
  ethical   55–6, 57, 179, 204 n11
staged drowning   5, 121, 133, 137, 150,
    186–7, 197 *see also* real drownings
Stockwell Training College   17, 34
suffrage movement   27, 38, 41, 46, 99,
    104, 108, 120, 174, 192, 209 n10,
    218 n11
Sydenham Road Girls School   34, 42, 190

Tait, Matron   157–60
teacher training   33, 36
teaching salary   31–2, 34, 190–2, 202 n12
technical education   110–14
Thornton Heath   11, 13, 16–17, 19, 24,
    26, 34, 53–5, 98, 190
Townley House Ladies School   30, 35
Trade Schools   5, 105, 107, 113, 115, 116,
    117–21, 211 n26
transport/travel
  car   146, 148
  difficulties of   72, 73, 74, 77–8, 98–9,
    139, 146, 147–8
  ease of   4, 9, 10–11, 182
  horse-drawn   4, 11, 26, 87, 142, 146,
    147, 148
  railway   4–5, 9, 72, 87, 91, 98–9, 111,
    148, 183
  steamer   4, 72, 139, 141, 145, 148

University College London   42, 191
utopian ideals   59, 140, 180, 182
utopian novels   50–1, 53 *see also* Morris,
    William *and* Wells, H. G.

Victorian lifestyle   4, 9–18, 25, 26,
    29, 39–40, 43–4, 46–51 *see*
    also education; marriage *and*
    respectability; *compare with*
    colonial life, in New Zealand
vocational training   20–1, 105

waistcoat-making   5, 105, 110,
    115–19
Webb, Beatrice (née Potter)   44, 50–1,
    100, 108, 141, 148, 204 n5, 208
    n3 *see also* Fabian Society; social
    reform *and* Women's Industrial
    Council
Webb, Sydney   44, 49, 50–1, 120–1, 205
    n5 *see also* Fabian Society *and*
    London County Council
Wellington   6, 133, 137, 139, 140, 141–5,
    148, 171, 173, 177, 184
Wells, H. G.   49, 51, 138, 162, 187, 204
    n5 *see also* Fabian Society
WIC *see* Women's Industrial Council

Wilks, Dr Elizabeth   192–3, 218 n13
    see also suffrage movement
Williams, Dr J. W.   155
WIN see Women's Industrial News
Women's Industrial Council   44, 47, 49,
    64, 97–100, 106–8, 110–12, 114,
    117–21, 140–1, 152, 171, 187
    see also London County Council
    and waistcoat-making
box-making   102, 103
flower-making   101–2, 104–5
investigation committee   100
education and technical training
    committee   113–14
Women's Industrial News   100, 107